# A little course in...
# Astronomy

A little course in...

# Astronomy

LONDON, NEW YORK, MUNICH,
MELBOURNE, DELHI

**Senior Editor** Helen Fewster
**Senior Art Editor** Alison Shackleton
**Managing Editor** Penny Warren
**Picture Researcher** Sarah Hopper
**Senior Jacket Creative** Nicola Powling
**Jacket Designer** Rosie Levine
**Senior Pre-production Producer** Tony Phipps
**Senior Producer** Alex Bell
**Art Director** Jane Bull
**Publisher** Mary Ling

**DK India**
**Editor** Ligi John
**Assistant Editor** Neha Ruth Samuel
**Senior Art Editor** Ira Sharma
**Art Editor** Anjan Dey
**Assistant Art Editor** Pallavi Kapur
**Managing Editor** Alicia Ingty
**Managing Art Editor** Navidita Thapa
**Pre-production Manager** Sunil Sharma
**DTP Designers** Rajdeep Singh, Neeraj Bhatia

**Written by** Robert Dinwiddie

First published in Great Britain in 2013 by
Dorling Kindersley Limited, 80 Strand, London WC2R 0RL
Penguin Random House (UK)

2 4 6 8 10 9 7 5 3 1
001–196196–Dec/2013

Copyright © 2013 Dorling Kindersley Limited

All rights reserved. No part of this publication may be
reproduced, stored in a retrieval system, or transmitted in any
form or by any means, electronic, mechanical, photocopying,
recording, or otherwise, without the prior written consent
of the copyright owners.

A CIP catalogue record for this book is available
from the British Library.

ISBN 978-1-4093-3980-9

Printed and bound by Leo Paper Products Ltd, China

Discover more at
**www.dk.com**

# Contents

Build Your Course 6 • Our place in Space 8 • Our place in the Universe 10 • Earth and its closest neighbours 12 • Gas Giants and Asteroids 14 • What can you see? 16 • Getting Started 22

## 1
## Start Simple

| | |
|---|---|
| The Moon and its phases | 26 |
| Observing a Lunar Eclipse | 32 |

**NAVIGATING THE NIGHT SKY**     **34**
How Stars Move • Constellations and Asterisms • The view from Earth • Locating the North and South Celestial Poles

**OBSERVATION TECHNIQUES**

| | |
|---|---|
| The Milky Way | 54 |
| Locating Stars and Constellations | 58 |
| Jupiter | 68 |
| Meteor Showers | 74 |
| Venus | 82 |
| Artificial Satellites | 86 |
| A Solar Eclipse | 90 |

**QUICK GUIDES**

| | |
|---|---|
| Stars | 50 |
| How Star Systems Work | 52 |
| The Milky Way Galaxy | 56 |
| The Zodiac | 66 |
| Jupiter | 70 |
| Meteors and Meteorites | 76 |
| Venus | 84 |
| Gravity and Orbits | 88 |

**KEY CONSTELLATIONS**

| | |
|---|---|
| Orion | 60 |
| Scorpius | 64 |
| Perseus | 72 |
| Cygnus | 78 |

**STARHOPPING**

| | |
|---|---|
| From the Plough | 46 |
| From Crux | 48 |
| From Orion | 62 |
| From the Summer Triangle | 80 |

## 2
## Build On It

| | |
|---|---|
| Choosing and using binoculars | 94 |

**OBSERVATION TECHNIQUES**

| | |
|---|---|
| The Moon through binoculars | 96 |
| Mars | 110 |
| The Orion Nebula | 114 |
| The Andromeda Galaxy | 128 |
| Mercury | 136 |
| A Comet | 140 |

**QUICK GUIDES**

| | |
|---|---|
| The Moon | 98 |
| Star Clusters | 102 |
| Mars | 112 |
| Light and Dark Nebulae | 116 |
| Galaxies | 130 |
| Mercury | 138 |
| Comets | 142 |

**KEY CONSTELLATIONS**

| | |
|---|---|
| Taurus | 100 |
| Cassiopeia | 104 |
| Hercules | 106 |
| Centaurus | 108 |
| Ophiuchus | 118 |
| Crux | 120 |
| Sagittarius | 122 |
| Carina | 124 |
| Andromeda | 126 |
| Leo | 132 |
| Dorado | 134 |

## 3
## Take It Further

| | |
|---|---|
| Choosing and using a small telescope | 146 |

**OBSERVATION TECHNIQUES**

| | |
|---|---|
| Saturn | 148 |
| Jupiter's Galilean Moons | 152 |
| Safely observing the Sun | 160 |
| A Supernova Remnant | 170 |

**QUICK GUIDES**

| | |
|---|---|
| Saturn | 150 |
| Planetary Nebulae | 156 |
| The Sun | 162 |
| Supernovas | 172 |

**KEY CONSTELLATIONS**

| | |
|---|---|
| Aquarius | 154 |
| Vela | 158 |
| Virgo | 164 |
| Ursa Major | 166 |
| Hydra | 168 |

**SKY MAPS**     **176**

**Index**     **188**
**Useful resources**     **191**
**Acknowledgements**     **192**

**PUBLISHER'S NOTE**
Looking at the Sun with the naked eye, binoculars, or a telescope can cause eye damage. Advice on safe viewing of the Sun is included on pages 160 and 161 of this book. The author and the publishers cannot accept any liability to readers failing to follow this advice. Modifying cameras or other equipment may invalidate the manufacturer's warranty and readers do so at their own risk.

# Build Your Course

This course is divided into three parts. Start Simple introduces the night sky, focusing on objects you can see with the naked eye. Build On It explains what you can see with binoculars, while in Take It Further you'll learn about objects best viewed through a telescope.

## Observation techniques and activities

Within each chapter you'll find information on particular astronomical objects or phenomena to look for in the night sky – such as the Moon, a specific planet, a bright comet, an eclipse, or a particular deep sky object. You'll also find advice on when and where you will be able to see these specific objects or events, as well as what to look out for when viewing them, and tips on observation technique and how to navigate your way across the sky at night.

**Taurus constellation figure**

## Key constellations

In each chapter, you'll also find many double-page features that focus on key constellations. Each includes a figure and star map for the constellation, with labels indicating the more interesting objects to view in it, and "locator" diagrams to help you work out when and where to observe the constellation in the night sky. Some of the constellations covered are visible to viewers in one hemisphere only (northern or southern), but many can be seen by viewers in either hemisphere.

**Taurus locator diagram**  **Crab Nebula in Taurus**

## How does it work?

In addition to the various observation exercises and constellation-based features, look out for background information – as well as concise explanations on how various astronomical phenomena work – in the "Quick Guides" scattered throughout the book. For example, there are quick guides to all the bright planets, the Sun and Moon, and to stars, comets, and the Milky Way Galaxy. You'll also find plenty of explanatory diagrams, covering everything from satellite orbits to the different types of solar eclipse.

*Diagrams and artworks help explain how things work*

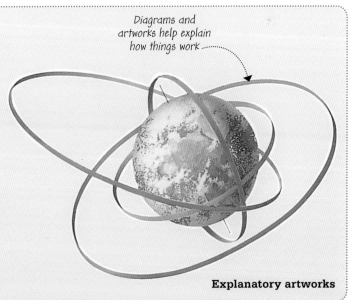

**Explanatory artworks**

INTRODUCTION

# Our place in Space

Our planet is travelling through a boundless, three-dimensional entity called "space". This is almost empty, but at varying distances from Earth vast numbers of other concentrations of matter – planets, stars, and galaxies – also exist. These objects, and the emptiness separating them, are known as "the Universe". Astronomy is the study of these objects.

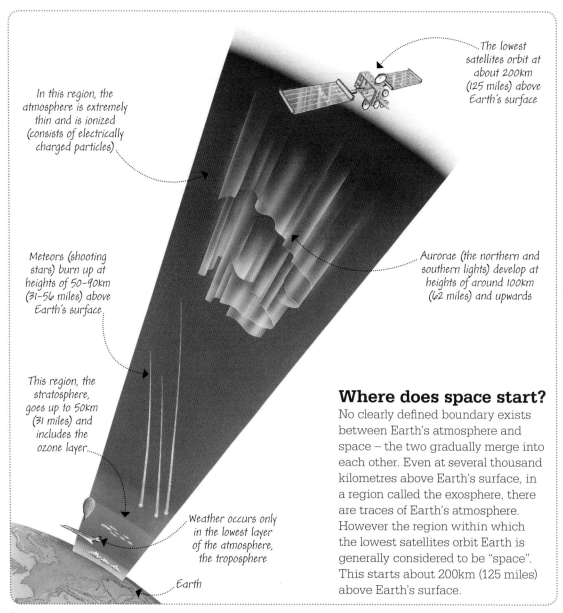

The lowest satellites orbit at about 200km (125 miles) above Earth's surface

In this region, the atmosphere is extremely thin and is ionized (consists of electrically charged particles)

Meteors (shooting stars) burn up at heights of 50-90km (31-56 miles) above Earth's surface

Aurorae (the northern and southern lights) develop at heights of around 100km (62 miles) and upwards

This region, the stratosphere, goes up to 50km (31 miles) and includes the ozone layer

Weather occurs only in the lowest layer of the atmosphere, the troposphere

Earth

## Where does space start?

No clearly defined boundary exists between Earth's atmosphere and space – the two gradually merge into each other. Even at several thousand kilometres above Earth's surface, in a region called the exosphere, there are traces of Earth's atmosphere. However the region within which the lowest satellites orbit Earth is generally considered to be "space". This starts about 200km (125 miles) above Earth's surface.

INTRODUCTION

# The big picture

The Universe is vast – in fact it is possibly infinite in size – and everything in it is part of something bigger. One way to get a feel for the size of the Universe is to use a series of astronomical stepping stones, as shown here – from our planet, to the Solar System that contains our planet, to the Milky Way galaxy, and so on. The largest known structures in the Universe are strings of galaxy superclusters, which extend for billions of light years through space.

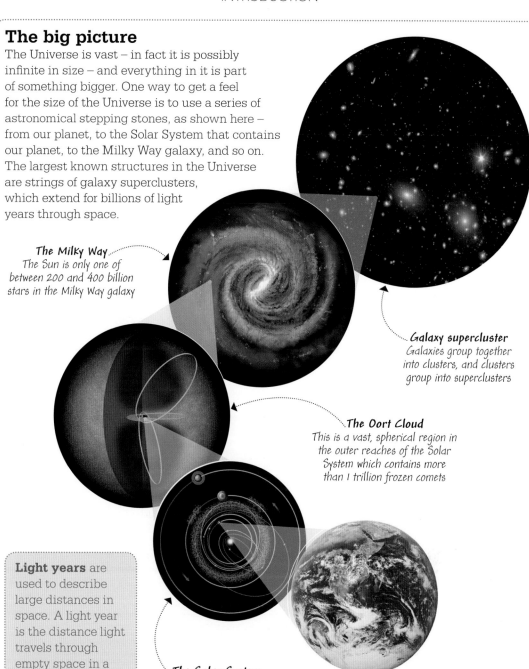

**The Milky Way**
The Sun is only one of between 200 and 400 billion stars in the Milky Way galaxy

**Galaxy supercluster**
Galaxies group together into clusters, and clusters group into superclusters

**The Oort Cloud**
This is a vast, spherical region in the outer reaches of the Solar System which contains more than 1 trillion frozen comets

**Light years** are used to describe large distances in space. A light year is the distance light travels through empty space in a year; it's just under 10 trillion km – or 6 trillion miles.

**The Solar System**
This consists of the Sun, together with eight planets and many smaller objects that orbit the Sun

**Planet Earth**
Earth is one of four rocky planets in the Solar System

INTRODUCTION

# Our place in the Universe

Our neighbourhood in space is the Solar System. This consists of the Sun, together with eight planets and many smaller objects – such as asteroids, comets, and planetary moons – that orbit the Sun.

## The Solar System

Objects in the Solar System fall into different groups, which generally occupy different regions. Four rocky planets (Mercury, Venus, Earth, and Mars) are closest to the Sun. Beyond Mars is the "main asteroid belt", where rocky asteroids orbit. Next is a region in which four much larger planets – "gas giants" Jupiter, Saturn, Uranus, and Neptune – move around. Beyond Neptune lie two vast areas occupied by comets and other small objects made of rock and ice: the Kuiper belt, and further out, the Oort cloud.

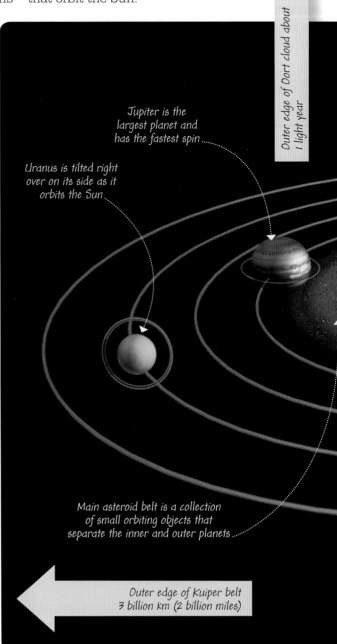

Jupiter is the largest planet and has the fastest spin

Uranus is tilted right over on its side as it orbits the Sun

Outer edge of Oort cloud about 1 light year

Main asteroid belt is a collection of small orbiting objects that separate the inner and outer planets

Outer edge of Kuiper belt 3 billion km (2 billion miles)

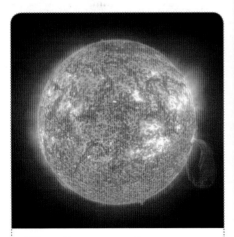

## The Sun our closest star

The Sun is a medium-sized star – a huge ball of hot, luminous gas, which produces energy from nuclear fusion reactions in its core (central region). It has existed in this form for some 4.6 billion years and should stay the same for another 5 billion years or so.

INTRODUCTION

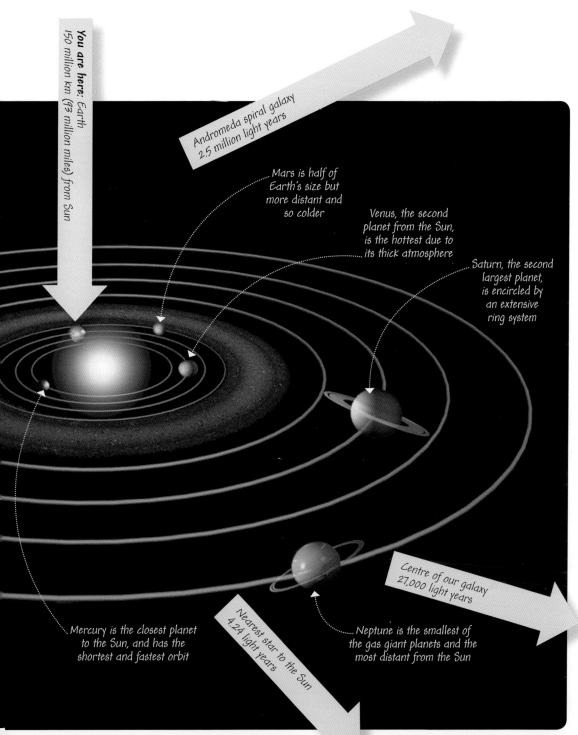

INTRODUCTION

# Earth and its closest neighbours

The Solar System's four inner planets – Earth, Mars, Venus, and Mercury – look very different on the surface, but they were formed from the same materials and share many features in common. All four have a central iron core surrounded by a mantle and crust of silicon-rich rock.

## The rocky planets

The inner planets are often described as "rocky" or "terrestrial" because they formed from rock and metal. As they developed, heavier materials like metal sank to the core, and lightweight rock rose to the surface. Sun-scorched Mercury has almost no atmosphere, but Venus and Mars are covered by a blanket of carbon dioxide. Uniquely Earth is the only planet with oxygen-rich air and liquid water on the surface.

**Size matters** Over time planets and objects roving across the sky appear to change in size. This apparent size, or angular diameter, is measured in degrees, arcminutes, or arcseconds: 1° is 60 arcminutes; 1 arcminute is 60 arcseconds. A closed fist held at arm's length is approximately 10°.

Venus's surface is covered by solidified lava from volcanic eruptions

Deep fractures run across Venus, forming winding canyons

Mercury's dusty surface is covered in impact craters from meteorites

Mercury          Venus

# INTRODUCTION

## Our natural satellite

Earth is the only rocky planet with a large natural satellite. The Moon formed around 4.5 billion years ago, coalescing from a cloud of debris after Earth was struck by a minor planet. The dark "seas", or maria, on the Moon's surface are plains of volcanic rock that formed when large impacts fractured the Moon's crust, releasing vast floods of lava onto the surface. Surrounding the sea are paler highlands covered in ancient craters. Like the rocky planets, the Moon is thought to have a partially molten core of iron. It is the brightest object in the sky after the Sun.

About three-quarters of Earth's surface is covered in oceans

Mars gets it reddish colour from iron oxide (rust)

Like all the rocky planets, Mars has a core of red-hot iron

Earth

Mars

# INTRODUCTION

# Gas Giants and Asteroids

Beyond the main asteroid belt lie four gigantic planets: gas giants Jupiter, Neptune, Saturn, and Uranus. Very different from the rocky inner planets, these strange worlds are vast gas and liquid globes composed of substances such as hydrogen, helium, and methane.

## The gas giant planets

Despite the name, the gas giants are mostly liquid and have small solid cores. All have a deep, stormy atmosphere and rings made of rock or ice fragments. Mighty Jupiter is the largest planet by far, its mass more than twice that of all other planets combined. Like Saturn, it spins so fast that its surface clouds have stretched out into bands.

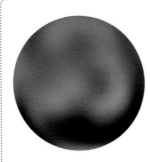

## Remember Pluto?

When Pluto was discovered in 1930, it was considered to be the ninth planet. In recent years astronomers have found numerous Pluto-like objects in the outer Solar System – some larger than Pluto. As a result, Pluto was reclassified in 2006 as a dwarf planet.

Jupiter's cloudy outer atmosphere is 1,000km (620 miles) thick

Deep inside Jupiter, intense pressure turns hydrogen into a liquid metal

A layer of liquid hydrogen and helium lies under the outer atmosphere

The Great Red Spot is a storm that has been raging for more than 300 years

Jupiter

Neptune

# INTRODUCTION

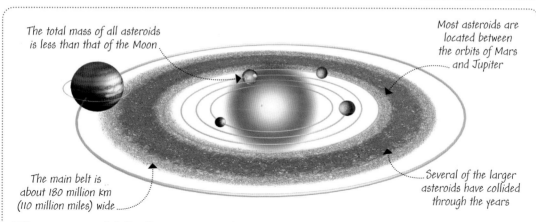

- The total mass of all asteroids is less than that of the Moon
- Most asteroids are located between the orbits of Mars and Jupiter
- The main belt is about 180 million km (110 million miles) wide
- Several of the larger asteroids have collided through the years

## The asteroid belt

Asteroids are giant rocks that drift around the inner Solar System. Most lie in a belt between Mars and Jupiter, but some occasionally come dangerously close to Earth. The smallest are the size of houses, while the very largest is also classified as a dwarf planet.

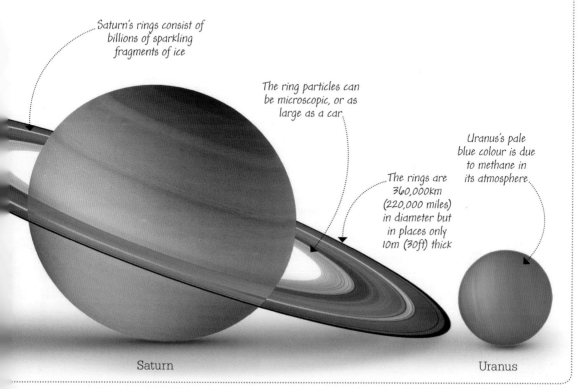

- Saturn's rings consist of billions of sparkling fragments of ice
- The ring particles can be microscopic, or as large as a car
- The rings are 360,000km (220,000 miles) in diameter but in places only 10m (30ft) thick
- Uranus's pale blue colour is due to methane in its atmosphere

Saturn

Uranus

INTRODUCTION

# What can you see?

With or without binoculars, there's plenty of "space" to view in the night sky. The Moon, planets, and the International Space Station all make regular appearances, but you can also see numerous stars and try to detect the patterns they make, known as constellations.

## One clear night

It's amazing what you can see in just one part of the night sky. This view to the south west, seen one March evening in Europe, has as its centrepiece the magnificent constellation of Orion. Above Orion and to its right are the planet Jupiter – one of several planets visible over the year – and two glittering star clusters, the Hyades and Pleiades. Not far from Orion is Sirius, the brightest star in the sky. With binoculars, once you know where to look you can see more types of objects, and even galaxies beyond our own.

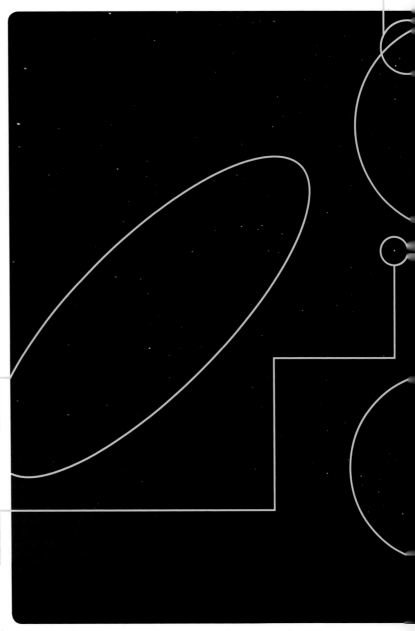

*Bright stars Castor and Pollux*

*Part of the constellation of Hydra*

*Bright star Procyon*

# INTRODUCTION

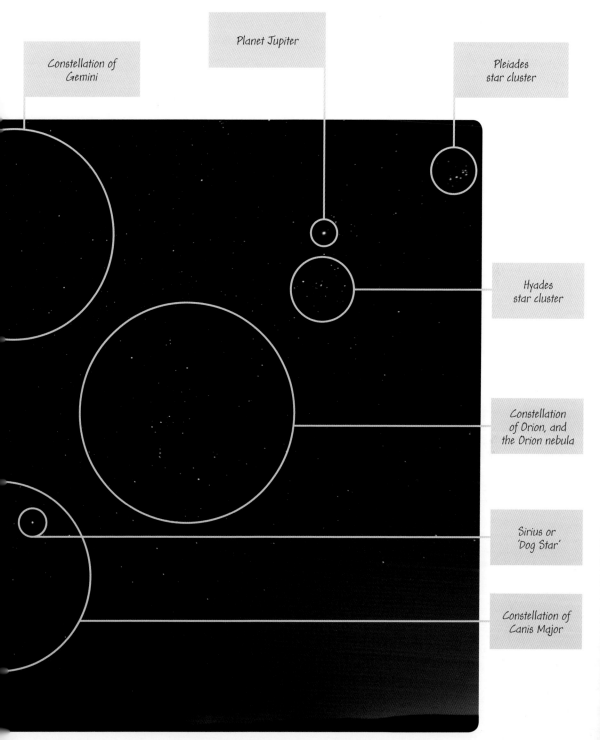

17

INTRODUCTION

# Occasional treats

Some astronomical phenomena, such as bright nebulae (glowing clouds of dust), can be seen on any clear night – provided you know where to look. But other events are much rarer, and you may have to wait for months or years to witness them. Some phenomena, such as eclipses, are predictable years or decades in advance, but the most spectacular comets tend to appear "out of the blue".

**Diary date?** Perhaps the ultimate rare astronomical event is a transit of Venus, when the planet Venus is seen to pass across the face of the Sun. The last transit happened in 2012; the next is due to occur in 2117.

## Showers of shooting stars

Meteors, or shooting stars, flash across the sky every night. But at various points in the year our planet encounters streams of dust that have come from bodies such as comets – and when this happens, meteor "showers" may be seen. Witnessing a shower is one of the easiest things you can do in astronomy: just lie back and look at the sky (see pp.74–75).

## Total solar eclipse
A total eclipse of the Sun is rarely seen from any fixed point on Earth: to see one, prepare to travel! Lunar and partial solar eclipses are more common (see pp.32 and 90).

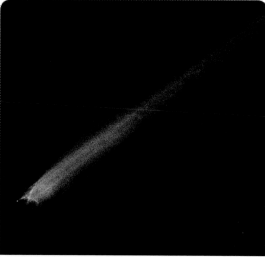

## Bright comets
Unpredictably and infrequently – about once every 5–10 years – a dazzling comet appears in the sky for a few weeks or even months. Try not to miss such an event (see p.140).

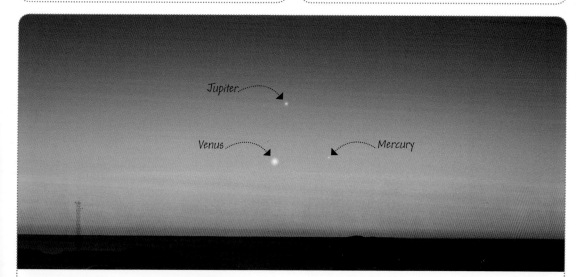

## Planetary conjunctions
Every now and again an impressive spectacle occurs when several of the bright planets – Venus, Jupiter, Mars, Mercury, and Saturn – appear close together in the sky and "dance" around each other over several nights. The dates of planetary conjunctions are worth seeking out and marking in your diary. Here, Jupiter, Mercury, and Venus conjoin at sunset.

INTRODUCTION

# What can you see **through a Telescope?**

A telescope can reveal a new level of detail in the night sky, but do be realistic about what you will see. With a small telescope some objects may still only be visible as pinpoints of light or tiny fuzzy discs, and even a large one is unlikely to reveal spectacular colours.

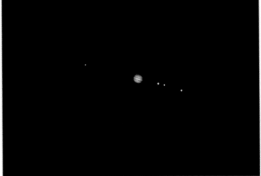

## Planetary moons

You don't need a telescope to see our Moon, but it will reveal a lot more surface detail. A small telescope may also allow you to view four of Jupiter's moons (see p.152), and some of Saturn's. The two images of Jupiter's moons shown above demonstrate how much more powerful the Hubble Space Telescope is compared with amateur equipment.

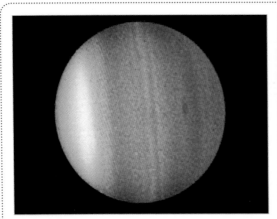
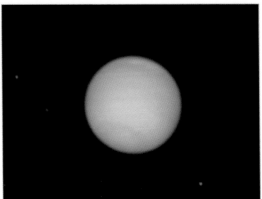

## Faraway planets

Occasionally Uranus (above, left) is just visible to the naked eye as a small pinpoint of light – but through a moderate-sized telescope you may be able to make out a small blue disc.

Still more remote is Neptune (above right); with a telescope you will probably only see a tiny blue dot. A large, powerful instrument is needed to make out any detail, let alone its moons.

# INTRODUCTION

## Distant galaxies

A few galaxies outside our own are visible to the naked eye, and more can be seen as fuzzy objects through binoculars. If you examine them with the more powerful optics offered by an amateur telescope you may be able to make out a bit more detail. This dramatic image of the Cigar Galaxy (M82) was made using a combination of space telescopes.

## What you won't see

**Black holes** No light ever escapes from these objects, so they are by definition invisible. Astronomers can detect black holes only by observing the effects they have on other nearby objects.

**Dark matter** This is thought to make up most of the matter in the Universe, but again by definition, it is invisible since it emits no light or other type of radiation. At the moment scientists are certain it exists but unclear exactly what it is.

**Exoplanets** These are planets orbiting stars other than the Sun. Astronomers have now found more than 900, but they can be seen – or more often detected – only with the most highly sophisticated scientific instruments.

INTRODUCTION

# Getting Started

The good news is that you don't need much in the way of special equipment to get started in astronomy. To avoid frustration, watch the weather forecast and pick a clear moonless night for your first outing – unless of course you're intending to observe the Moon! Seek out a good location from which to observe – for the best views choose a spot away from light pollution with a reasonably unobstructed view of the horizon.

## Make yourself comfortable

Bear in mind that you will probably be outside for long periods – often on cold nights – and looking up. It's important to wear appropriate clothing and to find a comfortable position. It takes about 20 minutes for your eyes to adapt to the dark, so you may need to be patient.

# INTRODUCTION

### Red light
A red LED light, or a torch covered with a red filter or cellophane, is a useful accessory. It will allow you to see what you are doing while your eyes remain adapted to the dark.

### Planning ahead
To help locate objects in the night sky, you may find it useful to acquire either a device called a planisphere or an astronomy "app" for smartphone or tablet (see pp.58–59).

### Binoculars or telescope
You can see a lot with the naked eye, but eventually you may want to invest in optical aids (see pp.94 and 146).

### Additional equipment
**Magnetic compass** to orientate yourself, unless you already have an accurate idea of the direction north or south, and can also identify other points on the compass.

**Warm comfortable clothing** including a warm windproof jacket, woolly hat, and fingerless gloves, unless you are observing from the tropics – in which case you may need mosquito repellent.

**Deckchair** for reclining comfortably as you observe; alternatively, a groundsheet and foam mattress are useful for lying flat.

**Notebook and pencil** for making notes and sketches of what you observe.

**Flask of coffee** to help keep you warm and awake.

# 1
# Start Simple

This chapter focuses on astronomy with the naked eye – concentrating on objects and phenomena in the night sky that you can observe without optical aids. These include the Moon, its movements and phases; the planets Jupiter and Venus; prominent stars and constellations such as Orion and the Southern Cross (Crux); meteor showers and shooting stars; and eclipses of both the Sun and Moon. You'll learn how to find your way around the night sky through "starhopping" exercises, and familiarize yourself with the concept of the celestial sphere and the use of a planisphere, which will help you to understand sky movements and locate objects in the night sky.

**The full Moon shortly after rising above the eastern horizon**

# Observing **the Moon**

Ask anyone to name a familiar object in the night sky, and the chances are they'll mention the Moon. It's visible for at least some of the time on most clear nights from anywhere in the world – and cannot be mistaken for anything else. But how well do you really know it? To kick-start your course in astronomy try keeping a lunar diary to learn about the Moon's movements.

### Tracking the Moon hour by hour

Choose a moonlit night and watch the Moon over the course of a few hours. You'll see that it moves slowly towards the western horizon, where it eventually sets. This daily movement occurs because Earth is constantly rotating on its axis. Earth takes 24 hours – a day – to complete its rotation; the overall effect is that the Moon appears to move in and out of sight.

# THE MOON

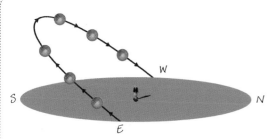

**Path as seen from northern hemisphere**

**Path as seen from southern hemisphere**

## One Moon, two hemispheres

The Moon rises in the east and sets in the west, but in between its movement differs depending on your location. Viewed from the northern hemisphere it traverses the southern sky; from the southern hemisphere it crosses the northern sky. Features on the surface also differ: in the southern hemisphere (as on p.26) it's upside down to northern hemisphere eyes.

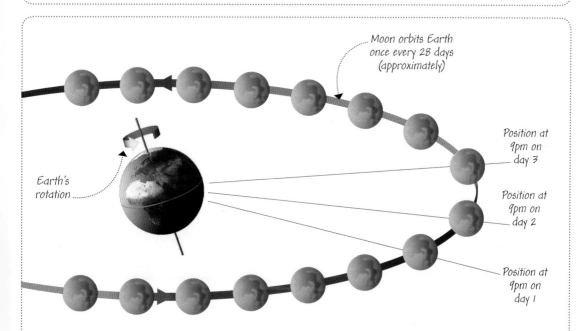

## Tracking the Moon day by day

Make a note of where the Moon is in the sky one evening at a particular time, say 9pm, and look for it again the next evening at exactly the same time. You'll find it is slightly to the east of where it was the first night – Earth's rotation has not quite brought it back to where it was 24 hours earlier. This shift to the east occurs every night and is caused by the fact that the Moon is orbiting Earth. Each 24 hours the Moon has moved round a bit in its orbit.

OBSERVATION TECHNIQUES

# Phases of **the Moon**

So far you've learned why the Moon moves across the sky from east to west – because Earth spins – and also that it displays a daily west-to-east shift as it orbits Earth. It's because of this orbit that the Moon's position in the sky relative to the Sun, as seen from Earth, changes – so over about a month the area of the Moon's surface that is lit up varies from zero to fully lit. The Moon's different appearances over this cycle are described as phases. During each cycle there are also changes in when, and for how long, the Moon can be seen each night.

## First half of the lunar cycle

During nights 0–14 of the lunar cycle (shown here as it appears in the northern hemisphere) on each successive night the sunlit area of the Moon gradually increases – it is said to be "waxing". Each night at sunset the Moon is further to the east, and can be seen in the sky for about 50 minutes longer before it sets.

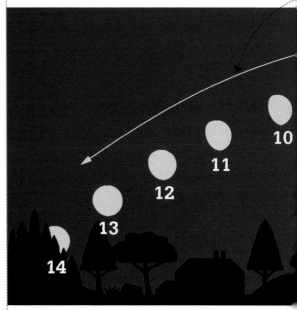

**THE SKY AT SUNSET Days 1–14**

**EAST** — Day 14: Full moon rises at sunset

### New Moon (Day 0)
In this phase the Moon lies between Earth and the Sun. The part of the Moon facing Earth is fully in shadow, and so for a few days it cannot be seen at all.

### Waxing crescent (Days 1–6)
A few days later the Moon has moved to a position where a thin sliver of its surface is lit up. In this phase the Moon is visible for a few hours around and after sunset.

28

# THE MOON

**Waxing gibbous**
During these phases the Moon is visible from late afternoon until the very early morning

**First quarter**
For northern hemisphere viewers, a first quarter Moon lies south at sunset

**Waxing crescent**
In these phases the Moon is quite close to the Sun in the sky, slightly restricting viewing chances

**LOOKING SOUTH** — Day 0: New Moon is invisible near the Sun — **WEST**

## First quarter (Day 7)
Seven days (about one quarter) into a lunar cycle, half of the Moon's face appears lit up. In this phase the Moon can be seen from the afternoon until about midnight.

## Waxing gibbous (Days 8–13)
The intermediate phase between the Moon being half- and fully lit is called "gibbous". A waxing gibbous Moon can be seen from late afternoon until the early morning hours.

OBSERVATION TECHNIQUES

**THE SKY AT SUNRISE Days 15–28**

*Waning crescent* — In these phases the Moon is close to the Sun, restricting viewing opportunities

*Third quarter* — For northern hemisphere viewers, the third quarter Moon lies south around sunrise

**EAST**  Day 29: New Moon is invisible near the Sun  **LOOKING SOUTH**

**Full Moon (Days 14–15)**
We see the Moon fully illuminated when it is on the opposite side of Earth from the Sun. In this phase it is visible all night, from sunset to sunrise.

**Waning gibbous (Days 16–21)**
The next phase after Full Moon is waning gibbous. In this phase the Moon rises between sunset and midnight and remains visible until it sets a few hours after sunrise.

# THE MOON

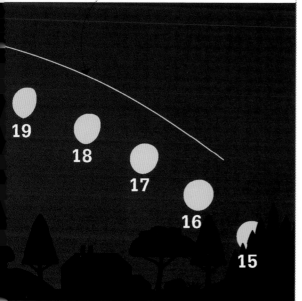

**Waning gibbous**
During these phases the Moon is visible from about 1–5 hours before midnight until after sunrise

Day 15: Full moon sets at sunrise    **WEST**

## Second half of the lunar cycle
In the second half of the lunar cycle comprising nights 15–28 (shown left, again for the northern hemisphere) on each successive night the area of the Moon's surface that is sunlit – as seen from Earth – gradually diminishes, and the Moon is said to be "waning". As its position in the sky shifts eastward a little every 24 hours, during this part of the cycle it becomes visible for about 50 minutes less each night, because it progressively becomes blotted out by the glare of the Sun.

## Southern hemisphere observers
If you are tracking the Moon from Australia, Brazil, or other parts of the southern hemisphere, things will look a little bit different. Although the changes that occur over the lunar cycle are exactly the same as they are in the northern hemisphere, you will see the "mirror image" of what a northern hemisphere viewer sees. For example, in its waning phases, from the southern hemisphere the illuminated part of the Moon appears on the right, whereas in the northern hemisphere it appears on the left.

### Third quarter (Day 22)
Three quarters of the way through a lunar cycle, the Moon's face is half lit up again. In this phase it rises at about midnight and is visible until noon.

### Waning crescent (Days 23–28)
In the waning crescent phase, the Moon rises a few hours before sunrise. Once the Sun has also risen it tends to fade from view quite quickly due to the Sun's glare.

# Observing a **Lunar Eclipse**

Once or twice a year from most parts of the world there's usually a chance to view a lunar eclipse. This can be a dramatic event, visible to the naked eye, in which Earth's full shadow covers the Moon; it occurs when the Sun, Earth, and Moon are aligned, with Earth in the middle. The table lists when and where you can see a total eclipse up to 2025.

## How a lunar eclipse works

During some of its orbits around Earth, the Moon passes through Earth's shadow. Like all shadows, it varies in intensity: full shadow is known as the Earth's "umbra"; part shadow is its "penumbra". A penumbral eclipse is when the Moon passes through Earth's penumbra, resulting in a slight dimming. A partial eclipse occurs when part of the Moon passes through Earth's umbra, and during a total eclipse the whole Moon passes through the umbra.

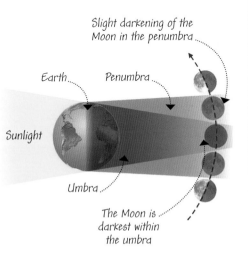

## Diary dates: total lunar eclipses

| Date | Location |
| --- | --- |
| 15 April 2014 | Australia, Pacific, Americas |
| 08 Oct 2014 | Asia, Australia, Pacific, Americas |
| 04 April 2015 | Asia, Australia, Pacific, Americas |
| 28 Sep 2015 | East Pacific, Americas, Europe, Africa, West Asia |
| 31 Jan 2018 | Asia, Australia, Pacific, Western North America |
| 27 Jul 2018 | South America, Europe, Africa, Asia, Australia |
| 21 Jan 2019 | Central Pacific, Americas, Europe, Africa |
| 26 May 2021 | East Asia, Australia, Pacific, Americas |
| 16 May 2022 | Americas, Europe, Africa |
| 08 Nov 2022 | Asia, Australia, Pacific, Americas |
| 14 Mar 2025 | Pacific, Americas, western Europe, West Africa |
| 07 Sep 2025 | Europe, Africa, Asia, Australia |

# LUNAR ECLIPSE

## Tracking a lunar eclipse

Lunar eclipses only occur when the Moon is full. The first sign is a slight darkening of the Moon as it passes into Earth's penumbra. During a penumbral eclipse – the least spectacular type – this is all that happens. In other types of eclipse, a much darker region moves across the face of the Moon as it passes into Earth's umbra (in this time-lapse photograph it can be seen happening from bottom left). During a total eclipse, which can take about an hour to develop, the Moon takes on a blood red appearance once the whole of it has passed into Earth's umbra (centre right in photograph). The reddening occurs because some sunlight is refracted towards the Moon by Earth's atmosphere. Subsequently the Moon slowly moves out of Earth's shadow.

**Note** About 2,350 years ago – when the world was believed to be flat – the Greek philosopher Aristotle put forward several arguments for believing that Earth was actually a sphere. One of his reasons was that the shadow cast upon the Moon in a lunar eclipse, has a curved edge.

# Understanding **How Stars Move**

So far you've tracked the Moon and learned how its phases and position in the sky relate to its journey around Earth, our planet's own rotation, and Earth's orbit round the Sun. Now it's time to investigate other objects in the night sky, such as stars and constellations; but first you need to understand a little more about how and why stars seem to move across the sky.

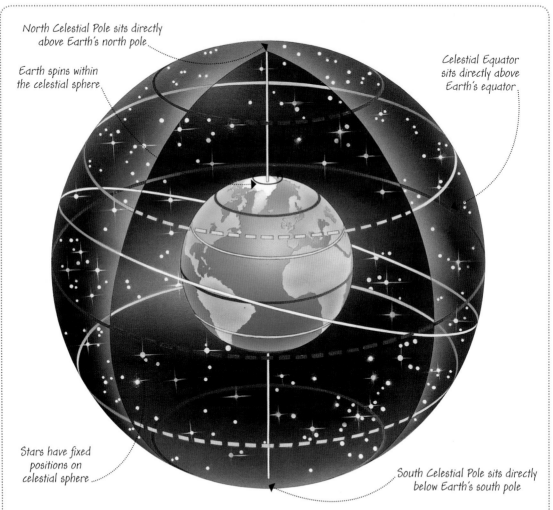

North Celestial Pole sits directly above Earth's north pole

Earth spins within the celestial sphere

Celestial Equator sits directly above Earth's equator

Stars have fixed positions on celestial sphere

South Celestial Pole sits directly below Earth's south pole

## The celestial sphere

Imagine a static shell-like sphere surrounding Earth in which all the distant stars are embedded at fixed positions on the inside shell. This imaginary globe is known as the "celestial sphere"; many of its features are "celestial" equivalents of those on Earth, including celestial north and south poles, and an equator (the celestial equator).

# Apparent star movement

Earth is spinning continuously within the celestial sphere. This means that when you look up at the night sky you see different patterns of star movement depending on your location – whether at the north pole, the south pole, the equator, or somewhere in between.

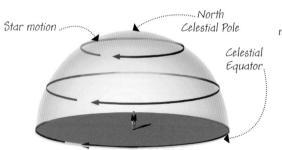

**Looking up from the north pole**
*Stars appear to circle anticlockwise around a point directly overhead – the north celestial pole. Stars at the horizon appear to travel around the horizon.*

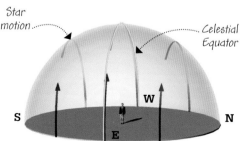

**Looking up from the equator**
*Stars appear to rise vertically in the east, swing overhead, and then drop vertically down again and set in the west.*

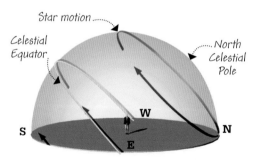

**At mid-latitudes (northern hemisphere)**
*Most stars rise in the east, cross the sky obliquely, and set in the west. But stars in the northernmost part of the sky move anticlockwise around the north celestial pole.*

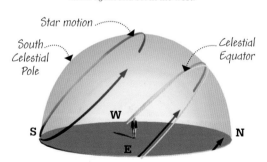

**At mid-latitudes (southern hemisphere)**
*Most stars rise in the east, cross the sky obliquely, and set in the west. But stars in the southernmost sky move clockwise around the south celestial pole.*

# Mapping the sky

The celestial sphere can be divided into six parts – two circular sections at the top and bottom centred on the celestial poles, and four sections around the middle, with the celestial equator running horizontally through. As stars and other distant objects have a fixed place on the celestial sphere, it becomes possible to produce maps of these sections showing their positions. You'll find two polar sky maps and four equatorial maps on pp.176–187.

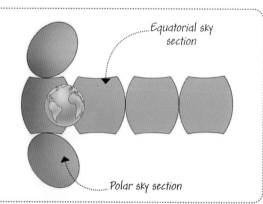

# Constellations and Asterisms

Since about 1930, astronomers have divided the celestial sphere into 88 areas, called constellations. An example of a constellation – Aquila, the Eagle – is shown below. It has an irregular border, and surrounding it are other constellations, such as Delphinus and Scutum.

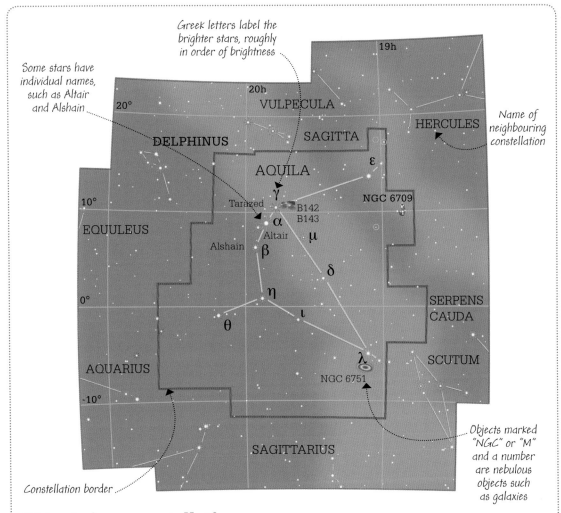

## Objects in a constellation

On a constellation map, the brighter stars are joined by lines to form what is called the constellation figure. Some individual stars are labelled. The oldest systematic method for naming stars uses Greek letters, with the brightest star in a constellation designated α, the second brightest with β, and so on. Other objects, such as galaxies, are also labelled.

# CONSTELLATIONS AND ASTERISMS

## Constellation figures

Every constellation has a figure – usually an animal (such as Aquila the Eagle), an object, or a mythological character – and many of them represent shapes seen among groups of stars. Often they originated with the ancient Greeks, predating the regions defined by recent astronomers. They are essentially imaginary: the lines joining the stars in a figure are not seen in the sky. You'll learn how to identify some of the better known constellations as well as the more prominent stars in those figures and other objects (galaxies and nebulae) within the constellation regions.

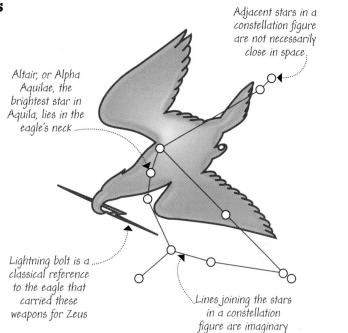

Adjacent stars in a constellation figure are not necessarily close in space

Altair, or Alpha Aquilae, the brightest star in Aquila, lies in the eagle's neck

Lightning bolt is a classical reference to the eagle that carried these weapons for Zeus

Lines joining the stars in a constellation figure are imaginary

## Asterisms

Within some constellations are distinctive groupings of stars called asterisms. Examples include Orion's Belt in the constellation of Orion (see pp.60–61), the Teapot asterism in Sagittarius, the Sickle asterism in Leo, and the Plough asterism in Ursa Major. You'll come across a number of these asterisms as you go through the book – they can be useful signposts helping you to find the rest of the constellation figure or other nearby sky objects.

**Teapot asterism**

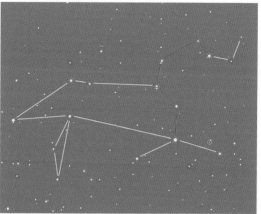

**Sickle asterism**

37

NAVIGATING THE NIGHT SKY

# The view **from Earth**

Earlier you learned how your position on Earth affects the pattern of star movement across the sky (see p.35). Your observing location – particularly your latitude – also affects what parts of the celestial sphere you are able to see.

## Effects of latitude

Your latitude indicates how far north or south of the equator you are observing from. The diagrams below show what you can see from three different positions: at the north or south pole, at the equator, and at a point about halfway between (mid-latitudes).

> **Remember** Your latitude affects the area of the sky that you can observe, but longitude makes no difference. So from New York City (USA), Madrid (Spain), and Beijing (China) – all at roughly the same latitude but very different longitudes – you'll observe more or less exactly the same areas of the night sky.

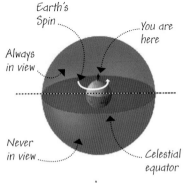

**Pole positions**
When you are at Earth's north pole you can only see the northern half of the celestial sphere – and it is always in view. The southern half of the sphere is never in view. If you're at Earth's south pole, then the reverse is true.

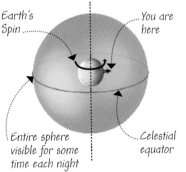

**Equatorial positions**
Standing at the equator, from an astronomical viewpoint you are in a much happier position: Earth's spin brings all parts of the celestial sphere into view for some of the time each night.

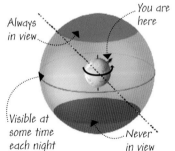

**Positions at mid-latitudes**
For mid-latitude locations, part of the celestial sphere is visible every night; part is sometimes in view; and a third part – a significant portion of the polar sky in the other hemisphere – will never be in view.

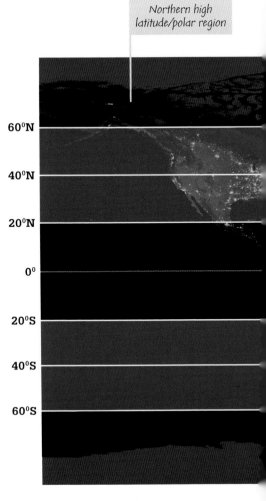

Northern high latitude/polar region

60°N
40°N
20°N
0°
20°S
40°S
60°S

THE VIEW FROM EARTH

## Constellation locators

Throughout the book you'll see "locators" – diagrams like those shown here. These are intended to help you find a constellation relative to the horizon at specific times and dates at different seasons of the year. They are standardized for observing positions at 40°N or 40°S. In most cases, if you are observing the night sky from a higher latitude than 40° (north or south), the constellation will appear correspondingly lower in the sky, and vice versa.

**1 April, 7pm
northern hemisphere**

**1 April, 7pm
southern hemisphere**

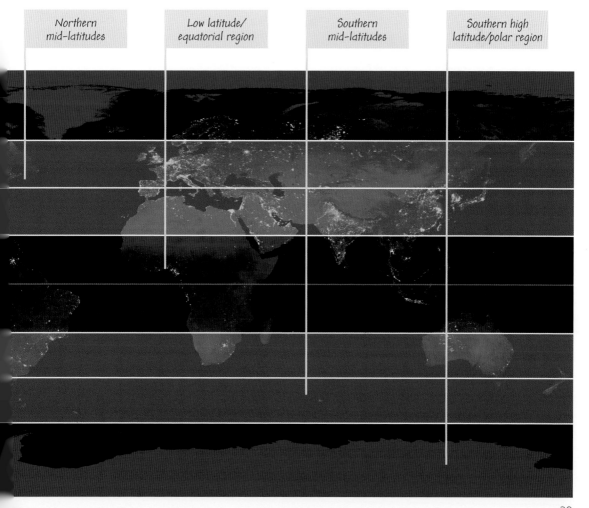

Northern mid-latitudes

Low latitude/equatorial region

Southern mid-latitudes

Southern high latitude/polar region

# Changes over **the Year**

You've seen how Earth's rotation brings different parts of the celestial sphere into view, but why does the night sky look different between the seasons? It's all to do with Earth's orbit of the Sun, which within a band called the zodiac (see pp.66–67) obstructs the direct view of various parts of the sky in turn.

## Shifting views

The part of the celestial sphere most easily viewed at any time of year is the region on the opposite side of Earth from the Sun. Month by month, this gradually shifts. Also in working out when to observe a particular part of the sky, there is a trade-off between date and clock time. For example, from a given location your view of the sky will be the same at midnight on 1 November; at 10pm on 1 December; and at 8pm on 1 January.

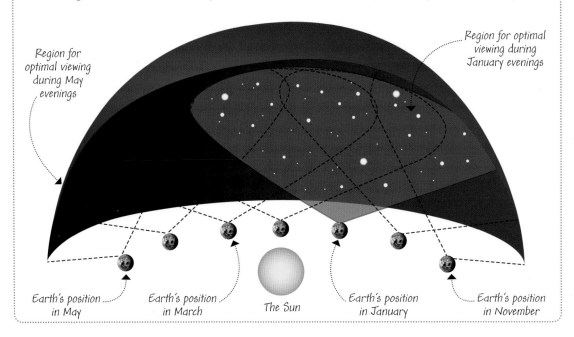

## Help is at hand!

If you're feeling confused about how your latitude, the time of year, and other factors such as clock time affect what you can see in the night sky – don't despair. A planisphere (right) is an astronomy aid that can help sort out these matters for you. You can learn how to use one – and find advice on some other types of astronomy aids – on pp.58–59.

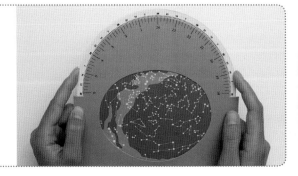

# Seasonal comparison

The two diagrams below compare what can be seen in the sky from northern hemisphere locations, looking south, in mid-March at 10pm and six months later in mid-September at 10pm. During the period between March and September, Earth travels halfway round its orbit of the Sun. This means that the view looking out at the celestial sphere completely changes, with different stars and constellations becoming prominent.

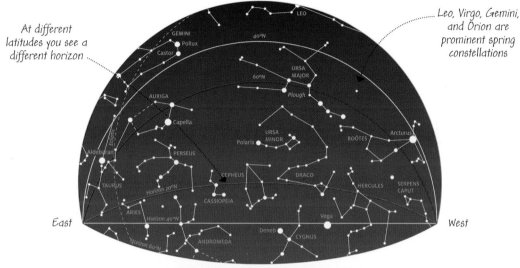

**Looking south in mid-March at 10pm**

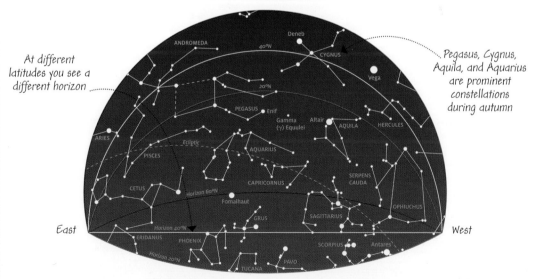

**Looking south in mid-September at 10pm**

# Navigating the Night Sky

The two points on the celestial sphere around which everything in the sky seems to continuously "wheel" due to Earth's rotation (see pp.34–35) are known as the north and south celestial poles. If you can find the appropriate pole for your hemisphere you will find it easier to navigate the night sky.

## Locating the North Celestial Pole

Conveniently a star called Polaris – part of a constellation called Ursa Minor (little bear) – is very close to the north celestial pole; at any specific location in the northern hemisphere, Polaris is always in the same place in the sky. Even better, two stars from a nearby asterism (star pattern) called the Plough point the way to Polaris.

**Tip** If you are observing from a low latitude – areas quite far south, near the equator – you may find it easier to locate the Plough at times of year other than autumn.

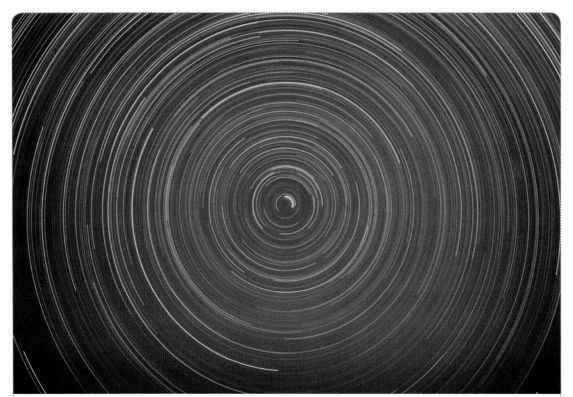

## Tracking Earth's rotation

If you observe Polaris, the Plough, and Ursa Minor over several hours one evening, you should just be able to see that these and all other stars in the sky are moving slowly in an anti-clockwise direction along circular paths with Polaris at the centre point. Just as it was with the Moon, this apparent movement of the stars is actually caused by Earth's rotation.

# LOCATING THE NORTH CELESTIAL POLE

## Finding the Plough

To locate the Plough on a clear autumn evening after dark, use a compass to find north, and look into the northern sky for an arrangement of stars that looks a bit like a saucepan. It should appear roughly as shown: the exact configuration depends on the time, and its height above the horizon depends on your latitude. Once you've found the Plough, use its "pointers", the stars Merak and Dubhe, to locate Polaris and Ursa Minor. The guide diagrams below help you find the Plough on winter, spring, or summer evenings: the procedure is the same, but the position of the Plough (and Ursa Minor) relative to the horizon will be different. A planisphere or tablet/smartphone "app" (see pp.58–59) may help. Having identified the Plough, use its pointers to locate Polaris.

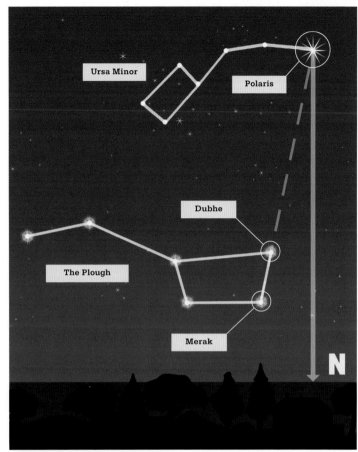

**Winter evenings**
1 Dec at midnight; also
1 Jan, 10pm; 1 Feb, 8pm

**Spring evenings**
1 March at midnight; also
1 April, 10pm; 1 May, 8pm

**Summer evenings**
1 June at midnight; also
1 July, 10pm; 1 Aug, 8pm

# Locating **Crux and the South Celestial Pole**

Unfortunately there is no bright star at the south celestial pole equivalent to Polaris in the northern hemisphere. However a spectacular constellation called Crux, or the Southern Cross, lies near to the south celestial pole. Crux is a stunning sight in itself, but also serves as a useful "pointer" to find other interesting objects in the southern night sky (see pp.48–49 and 120–121).

## Identifying Crux

Crux is a small constellation containing four bright stars in the shape of a cross. Three of the stars are appreciably brighter than the fourth. It can, however, be confused with a nearby pattern of four stars called the False Cross asterism. Crux is more compact than the False Cross, and also contains a fifth, moderately bright star within its outline.

Crux

False Cross asterism

## Tracking the stars

Once you've found Crux and identified the approximate location of the south celestial pole, try observing the whole southern night sky over an hour or so. You should be able to see that all the stars move slowly in a clockwise direction along circular paths, with the centre point of all the circles being the south celestial pole. If you were to take an extended-exposure photograph, the result would look something like that shown here. As with the Moon and Polaris, the apparent movement of the stars is actually due to Earth's rotation within the celestial sphere.

# LOCATING THE SOUTH CELESTIAL POLE

## Finding Crux

To locate Crux on a clear early winter evening after dark, pick a good observing location and use a compass to find south. Look into the southern sky for the arrangement of four bright stars in the shape of a cross. It should appear more or less as shown, but the exact configuration will depend on the time, date, and latitude from which you're observing. The two stars at either end of the long axis of the cross, Gacrux and Acrux, point towards the south celestial pole. To get there, follow a line through the long axis of the Cross for about three "cross lengths" until you reach the relatively empty region of the night sky containing the south celestial pole. This is the point around which everything in the southern sky seems to "wheel around" (see opposite).

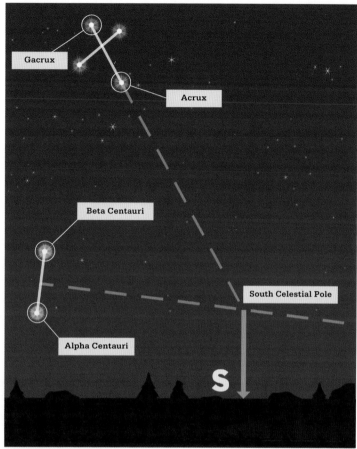

**Spring evenings**
1 Aug at 11pm; also
1 Sep, 9pm; 1 Oct, 7pm

**Summer evenings**
1 Nov at 2am; also
1 Dec, midnight; 1 Jan, 10pm

**Autumn evenings**
1 Feb at 11pm; also
1 Mar, 9pm; 1 April, 7pm

# Starhopping from the Plough

The Plough (the star pattern in Ursa Major that you identified on p.42) provides a signpost to several stars and constellations. Using a star pattern to navigate to neighbouring objects is called "starhopping", and it's a superb way to get familiar with the night sky. On a clear night, try these "star hops" from the Plough:

**1 To Polaris in Ursa Minor** First, as a reminder, find Polaris, the north polar star. Visualize a line between the last two stars in the Plough's "bowl", Merak and Dubhe. Extending the line past Dubhe brings you to Polaris.

**2 To Arcturus in Boötes, the Herdsman** Now visualize a curved line connecting the last three stars in the Plough's "handle", and extend this curve until you reach an exceptionally bright star with a golden hue. This is Arcturus, the leading star in Boötes. Once you've found Arcturus, you should be able to make out the main part of the constellation figure of Boötes.

**3 To Regulus in Leo** Starting at Megrez, the faintest of the seven stars in the Plough, draw an imaginary line to the star Phecda at the base of the Plough's "bowl". Continuing this line, you will pass the constellation of Leo Minor, before reaching the bright white star Regulus, in Leo. Once you've found Regulus, you should be able to make out the constellation figure of Leo – a crouching lion.

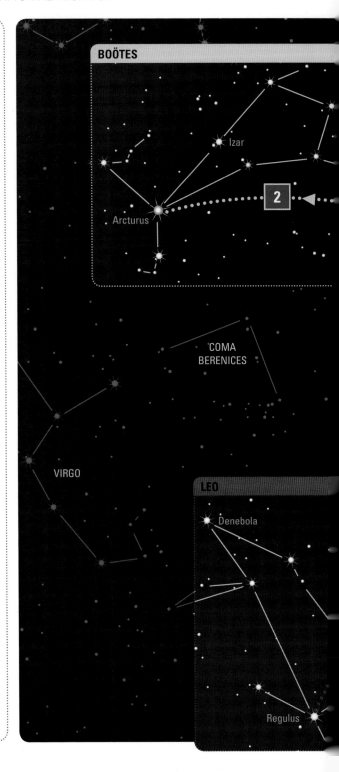

# STARHOPPING FROM THE PLOUGH

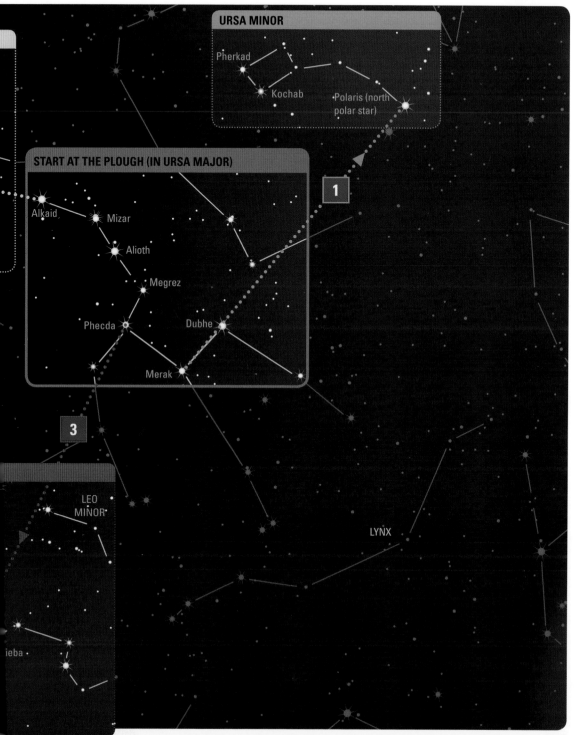

NAVIGATING THE NIGHT SKY

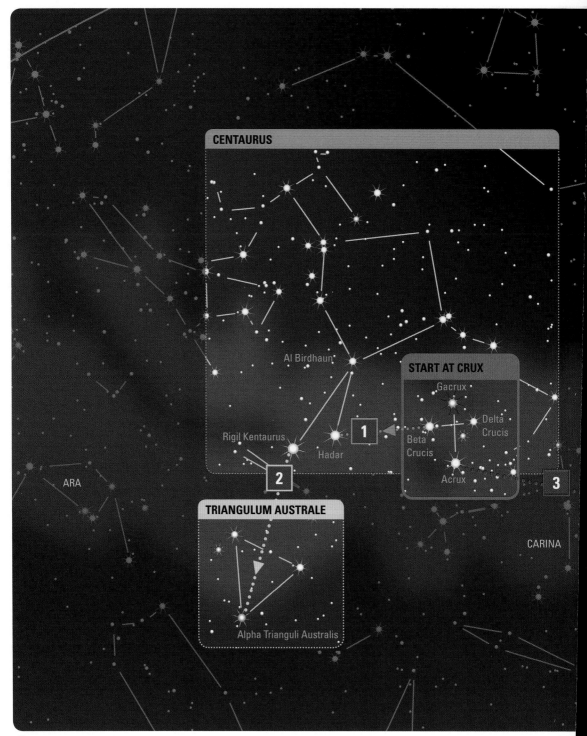

# Starhopping from Crux

Crux is the southern hemisphere's equivalent of the Plough – an excellent signpost for finding not just the sky's south pole, but also some neighbouring constellations and asterisms. Try these "star hops":

**1 To Centaurus** Start at Delta Crucis, the fainter of the two stars on the beam of Crux. Visualize a line through Beta Crucis, the other star on the beam. You'll soon reach the bright star Hadar. Just past this is an even brighter star called Rigil Kentaurus or Alpha Centauri. These are the two leading stars in the constellation of Centaurus.

**2 To Triangulum Australe** Use the chart to find a star in Centaurus called Al Birdhaun, or Epsilon Centauri. Extend an imaginary line from this through Rigil Kentaurus; the next bright star you come to is Alpha Trianguli Australis, or Atria. This forms one vertex of a prominent triangular constellation, Triangulum Australe. The other two vertices sit equally spaced in the direction back towards Centaurus.

**3 To the False Cross asterism** Imagine a line running from Hadar in Centaurus through Acrux – the brightest star in Crux – and extend it over the Milky Way. At about twice the distance between Hadar and Acrux, you'll reach one of the stars on the False Cross's "beam", Iota Carinae. You should then be able to spot the other three stars of the False Cross.

# A Quick Guide to **Stars**

Stars form from the condensation of clumps of gas and dust in galaxies: the stars that we see are part of our local galaxy, the Milky Way. Each star is an extremely hot ball of plasma (ionized gas) that generates energy through nuclear fusion reactions in its core (centre). Stars often occur in pairs or small groups.

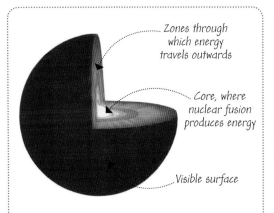

## Structure
A star consists of an energy-producing core – which in some stars has many layers – and surrounding zones through which the energy slowly moves outwards.

## Stellar variety
All the stars in the night sky appear to us as tiny pinpricks of light, but in fact stars vary enormously in their size, colour, and brightness, and also in their likely lifespans.

## Star size and colour
When stars are born, the mass of gas that forms them varies enormously. The star's mass defines not just its size but also its temperature – bigger stars are hotter – and lifespan – small stars tend to live longer. The temperature of a star, in turn, governs its colour: the hottest stars are blue-white, and the coolest are red. All this means that stars come in a wide variety of types, a few of which are shown here in comparison with the Sun.

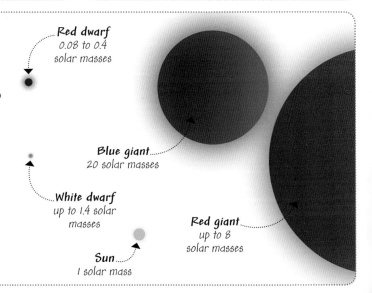

# Measuring magnitude (brightness)

Astronomers use two separate scales to express the brightness of stars and other sky objects. Apparent magnitude is a measure of how bright an object looks from Earth, while absolute magnitude expresses its true brightness, as if viewed from a standard distance. Amateur astronomers are most interested in apparent magnitude: faint objects are given high values on the scale, while bright objects have low or negative values.

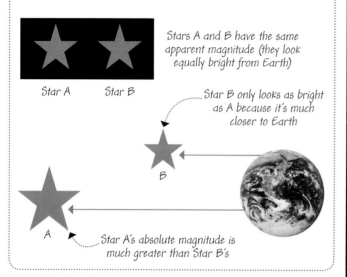

Stars A and B have the same apparent magnitude (they look equally bright from Earth)

Star B only looks as bright as A because it's much closer to Earth

Star A's absolute magnitude is much greater than Star B's

## Examples of apparent magnitudes

| Object | Maximum apparent magnitudes |
|---|---|
| The Sun | -26.7 |
| The Moon | -12.9 |
| Venus | -4.6 |
| Jupiter | -2.9 |
| Sirius (brightest nighttime star) | -1.4 |
| Canopus (second brightest nighttime star) | -0.72 |
| Alpha Centauri (star) | -0.27 |
| Vega (star) | 0.03 |
| Rigel (star) | 0.12 |
| Polaris (star) | 1.98 |
| Ganymede (moon of Jupiter) | 4.4 |

Rigel (in Orion)

STAR SYSTEMS

# How **Star Systems Work**

Our Sun is a solitary star, but the majority of stars in our galaxy have nearby companion stars and form binary systems of two stars, or multiple systems of three or more stars. Over time, some stars fluctuate in brightness – these are known as variable stars.

## Binary and multiple stars

In binary and multiple star systems, all the stars orbit a common centre of mass (or centre of gravity). Various types of orbital pattern may occur, some of which are shown below. These systems of gravitationally bound stars are not to be confused with pairs of stars – known as optical doubles – that appear to be close in the sky but are actually widely separated in space.

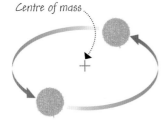
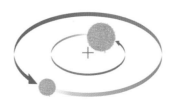
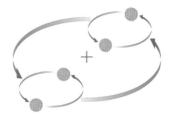

**Equal mass**
In binary systems where the two stars are of equal mass, the common centre of mass lies midway between the two stars.

**Unequal mass**
If one star in a binary system is more massive than the other, the centre of mass lies closer to the higher-mass star.

**Double binary**
In a double binary star system, each star orbits a companion, and the two pairs also orbit a single centre of mass.

## Eight-star system

Embedded in the Orion nebula (see pp.114–115) are eight stars that are collectively known as the Trapezium Cluster, because the four brightest – visible here with pink haloes – form a trapezium shape. Some of these stars are thought to be gravitationally bound and several are themselves binary stars, overall forming a complex system.

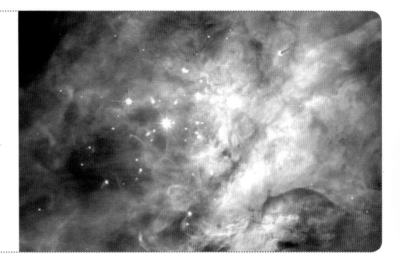

52

HOW STAR SYSTEMS WORK

## Variable stars

There are several reasons why the brightness of stars may vary. Occasional dips occur as one star in a binary system moves in front of the other (see Eclipsing binaries, below). Some stars have surface eruptions that make them brighter. Others, called pulsating variables, repeatedly puff up and contract, causing them to cyclically change in brightness.

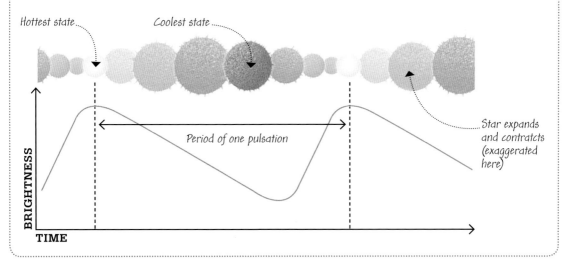

## Eclipsing binaries

In binary star systems, the overall brightness of the pair of stars may intermittently dip whenever one star moves in front of another. From the viewpoint of an observer on Earth, the star in the background is eclipsed (some light from it is partially cut off) so there is some dimming. Typically, there is a slight dimming when the fainter star in a pair is eclipsed and a more significant dimming when the brighter star is eclipsed.

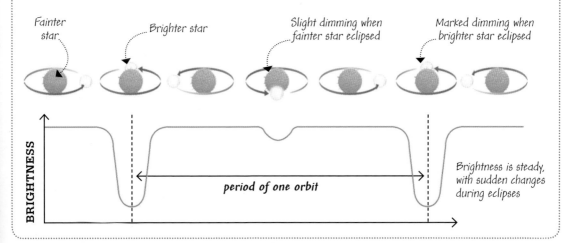

OBSERVATION TECHNIQUES

# Observing **the Milky Way**

On a clear night, if there is no Moon, and you can get well away from the light pollution created by towns and cities, there is a good chance you will be able to observe the Milky Way – the combined light of the vast majority of stars that make up the galaxy we live in. It is known as the Milky Way Galaxy because of the way it appears to us in the night sky.

## An arc of light across the sky

The Milky Way is a swathe of faint, milky-looking light that arcs right across the sky, from one horizon to the other. It's easily mistaken for a long cloud, but unlike a cloud it does not move relative to the individual bright stars you see in the sky. Depending on your latitude, and the time and date, the band of light may or may not pass directly overhead – for example, at some times or places it may arc across the northern or southern sky.

## Mapping the Milky Way

The Milky Way occupies a fixed area of the celestial sphere (see p.34): on maps it's shown as an irregular light-coloured band (see below and pp.176–187). Constellation figures are also mapped; some, such as Crux and Cassiopeia, are located along the Milky Way, whilst others – like Ursa Minor and Boötes – are not. This can help locate and identify constellations.

**Remember** All the individual bright stars you can see in the night sky with the naked eye are part of the Milky Way galaxy – even though when you look at a map many of them appear to be located outside the light band of the Milky Way. The reason for this is that most individual stars we can see are relatively close – within our local region of the Milky Way – and these nearby stars surround us in all directions.

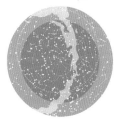 

**Northern hemisphere**     **Southern hemisphere**

# THE MILKY WAY

A section of the Milky Way, partly obscured by dark nebulae (clouds of dust)

# A Quick Guide to the Milky Way

The Milky Way is a vast whirlpool of at least 200 billion stars held together by gravity; it's around 100,000 light years across. In shape it resembles a pair of fried eggs held back to back, with a central bulge and a surrounding flat disc. Most of the stars are in the central bulge. Curving around this are two vast spiral arms and several smaller arms. Our Solar System lies in a minor arm known as the Orion Spur.

## Galactic centre

The white region in this image is Sagittarius A, a complex radio source thought to be the very centre of the Milky Way. Astronomers believe that a supermassive black hole lies hidden in this region. If stars or other material are sucked into this cosmic plughole by gravity, they disappear forever.

## The structure of the Milky Way

We may never get a view of our galaxy from far enough away to see its overall structure. However, by studying the distribution of stars in the Milky Way, astronomers are beginning to map out its exact shape. This diagram shows our current understanding of the Milky Way's geography.

Far 3 kiloparsec Arm

Sagittarius Dwarf Galaxy

Carina-Sagittarius Arm

Younger stars

Galactic Centre
In the middle of the central hub lies a turbulent region of gas clouds and older, heavyweight stars around a supermassive black hole

Perseus Arm

Dense molecular clouds

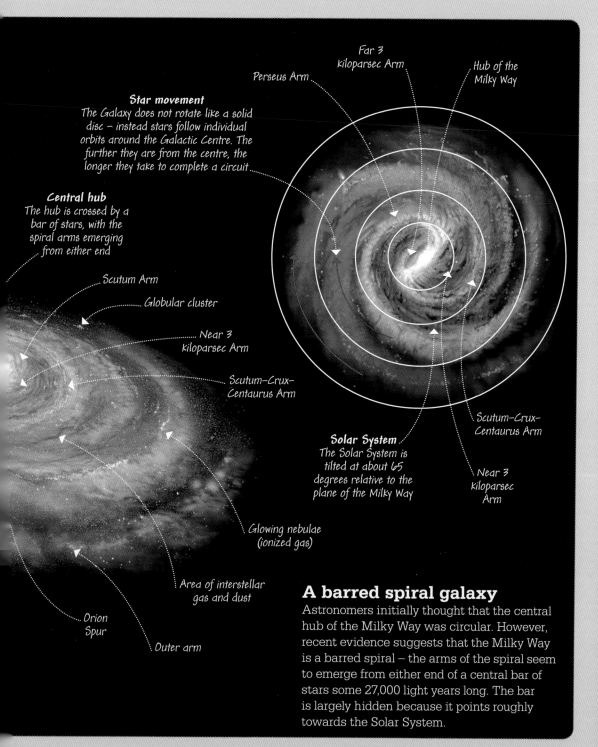

# A barred spiral galaxy

Astronomers initially thought that the central hub of the Milky Way was circular. However, recent evidence suggests that the Milky Way is a barred spiral – the arms of the spiral seem to emerge from either end of a central bar of stars some 27,000 light years long. The bar is largely hidden because it points roughly towards the Solar System.

TOOLS AND EQUIPMENT

# Locating **Stars and Constellations**

On many pages in this book you'll find locator diagrams that provide quick guidance on when, and where in the sky, you can see certain constellations, and the stars and other objects that they contain. But for more detailed guidance you may find it helpful to master using a planisphere, a smartphone or tablet astronomy "app", or other planetarium-style software.

## Get your bearings

You've seen that because Earth spins on its axis and moves around the Sun our view of the stars from Earth's surface is always changing. So, the first thing to do before using astronomical aids is to check the date, the time, your latitude, and compass directions at the site from which you want to observe. Armed with this information you are ready to use a planisphere, star chart or computer software to find out what should be visible – and where – at the time you want to observe the night sky.

## Smartphone and tablet apps

Over the past 20 years or so, dedicated astronomical software has been developed, based on the principle of the planisphere. Numerous packages are available – from simple, inexpensive "apps" for smartphones and tablets that plot the positions of stars, constellations, planets, and other objects for a particular time and place, to detailed programs with vast catalogues of stars and deep-sky objects. A number of these programs allow you to switch to red night-vision mode for use outside in the dark.

**Tip** Some recommendations for astronomy software are given on p.191. To start with, choose a simple "app" and play with it in daylight so that you are familiar with its features and ready to use it after dark.

# LOCATING STARS AND CONSTELLATIONS

## What is a planisphere?

A planisphere is a circular device that displays what is visible in the night sky at a particular date and time from a stated latitude. The map on the base layer is specific to that latitude, and is only suitable for observers in a band of about 5° to the north and south of it – so when choosing a planisphere it's important to check that it is appropriate for your observing location. They are available for most latitude bands except for the sparsely populated polar regions (see p.191 for recommendations).

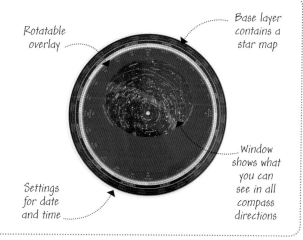

Rotatable overlay

Base layer contains a star map

Window shows what you can see in all compass directions

Settings for date and time

## Using a planisphere

To set the planisphere, turn the overlay until the time and date are aligned at the edge. Hold the planisphere up and turn the whole device so that the horizons marked at the edge of the window match the direction in which you want to look. The constellations and stars closest to that edge are what you see nearest the relevant horizon. Planispheres can also show when to see a particular constellation or star from your observing location. Find the constellation or star on the base map, rotate the overlay until you see it in the window, ideally away from any horizon, then look at the edge for a range of dates and times.

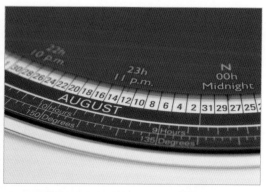

**Planisphere set for 11pm on 12 August**

# KEY CONSTELLATIONS

## Observing **Orion**

Orion, the Hunter, is arguably the most dramatic constellation in the sky – and a great starting point for beginners. It can be seen from both hemispheres and contains some glorious-looking objects.

## Orion, the Hunter

In the classic constellation figure, Orion holds a club in one hand and a shield in the other. A glittering trio of stars forms Orion's Belt. Around the belt are four more bright stars forming the figure's distinctive hourglass shape, including the red supergiant star Betelgeuse, and the blue-white supergiant Rigel. Below Orion's Belt is the "sword" region, which contains a nebula that looks spectacular through binoculars (see pp.114–115).

Betelgeuse, at Orion's right shoulder, has a diameter about 1,000 times that of the Sun

The six stars that form Orion's shield are collectively known as Pi Orionist

Orion's Belt is made up of Mintaka at the top, Alnilam ("string of pearls") in the centre, and Alnitak, lowest and to the left

Orion's sword includes the stunning Orion Nebula

Rigel, the brightest star in Orion, is about 40,000 times more luminous than the Sun

**Remember** Orion is most easily observed from November to March. It is difficult or impossible to observe Orion from either hemisphere between May and July because at this time it is too close to the Sun.

### **Northern hemisphere** locators

For observers at mid-latitudes (around 40°N), Orion becomes visible from early August in the south-eastern sky – but only in the early morning hours. By December it is prominent high in the south in the late evening.

**1 August, 5am**
Looking southeast

**1 December, midnight**
Looking south

**1 April, 7pm**
Looking southwest

# ORION

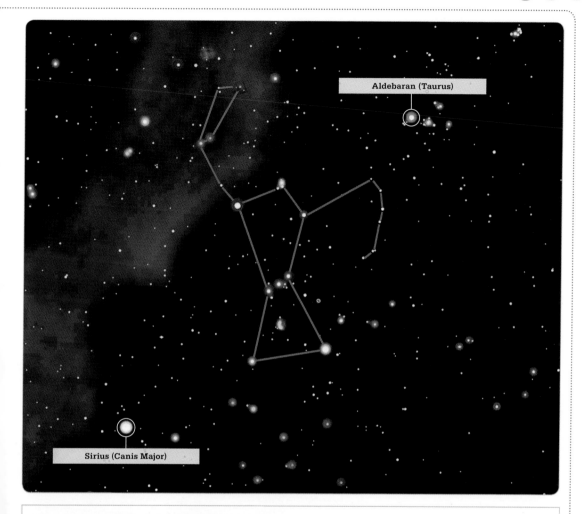

## Southern hemisphere locators

From mid-latitudes (around 40°S) Orion is visible during early mornings in August. By December it is prominent high in the north at around midnight – upside down compared to its appearance in the northern hemisphere.

**1 August, 5am**
Looking northeast

**1 December, midnight**
Looking north

**1 April, 7pm**
Looking northwest

# KEY CONSTELLATIONS

## Starhopping from Orion

Orion is a fantastic starting point for some "starhopping" exercises to identify nearby bright stars and the constellations in which they reside.

**1 To Sirius in Canis Major** Draw an imaginary line through Orion's Belt and extend it out from the same side of Orion on which Betelgeuse, the red supergiant, lies. You cannot possibly avoid stumbling upon Sirius, a dazzling pure white star that is the brightest in the entire night sky. Sirius is the leading star in the constellation of Canis Major, the Great Dog.

**2 To Aldebaran in Taurus** Starting again at Orion's Belt, visualize a line extending in the opposite direction, passing the star Bellatrix, which represents Orion's left shoulder. The destination is the bright red star Aldebaran – the "eye" of Taurus, the Bull. Continuing the line roughly onwards, you will reach a wonderful open star cluster, the Pleiades.

**3 To Procyon in Canis Minor** From Bellatrix, draw an imaginary line through Betelgeuse, on Orion's other shoulder. If you follow this for just over three times the distance between Bellatrix and Betelgeuse, you should arrive at another brilliant star, yellowy-white in colour, called Procyon. This is the leading star in the small and simple constellation of Canis Minor, the Little Dog. Note that Procyon, Sirius, and Betelgeuse form a large equilateral triangle.

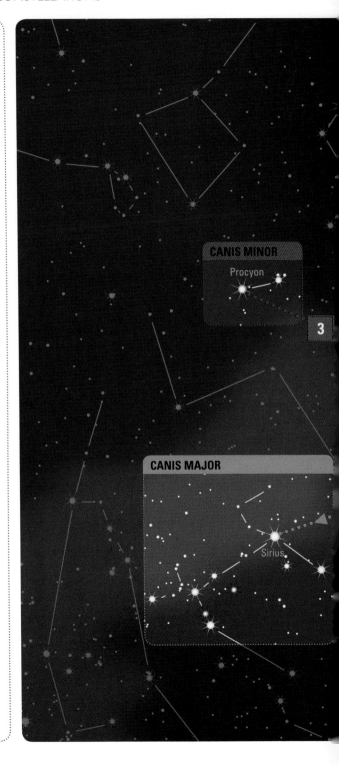

# STARHOPPING FROM ORION

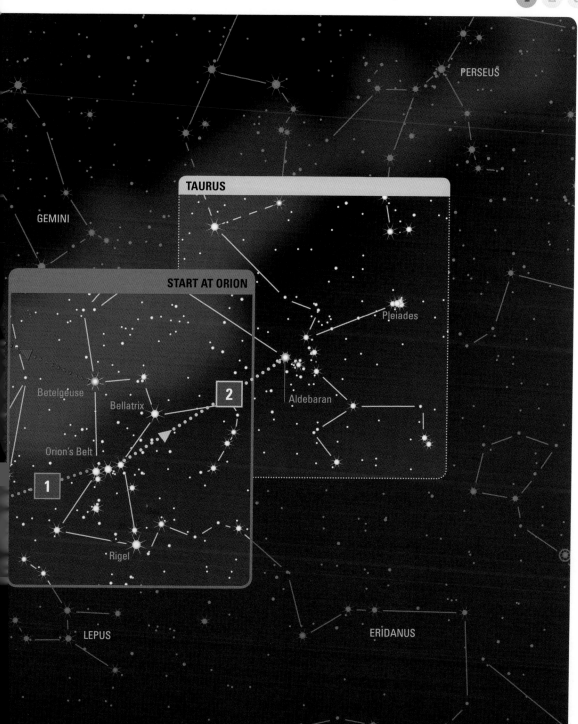

# KEY CONSTELLATIONS

## Observing **Scorpius**

The spectacular constellation of Scorpius represents the scorpion that in Greek mythology was sent by the goddess Artemis to kill Orion. It is most easily viewed from the southern hemisphere, but parts or even all of it may also be seen from most northern hemisphere locations.

> **Remember** The scorpion's tail passes through an area of the Milky Way that is rich in beautiful star clusters. Although to begin with you can enjoy looking at this stunning constellation with the naked eye, it's worth returning to later with binoculars.

### Scorpius, the Scorpion

Scorpius is easy to recognize. The heart of the beast is marked by the supergiant red star Antares, which is hundreds of times larger than the Sun and lies 600 light years away. A distinctive curve of stars passing across the Milky Way represents the tail – poised ready to strike. Further bright stars represent the scorpion's head and claws.

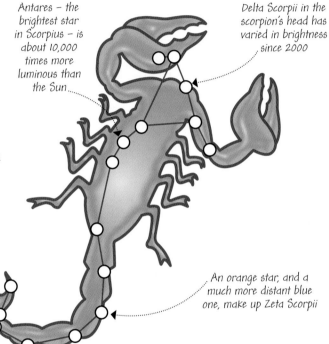

Antares – the brightest star in Scorpius – is about 10,000 times more luminous than the Sun

Delta Scorpii in the scorpion's head has varied in brightness since 2000

An orange star, and a much more distant blue one, make up Zeta Scorpii

Four moderately bright stars make up the sting in the tail of the scorpion

## **Northern hemisphere** locators

For observers at mid-latitudes, the main possibilities for viewing Scorpius occur between April and August, looking towards the southern horizon. In June nearly all of the constellation is above the horizon around midnight.

**1 April, 2am**
Looking south

**1 June, midnight**
Looking south

**1 August, 10pm**
Looking south

# SCORPIUS

## Southern hemisphere locators

From mid-latitudes Scorpius is visible from February in the early hours of the morning, looking east. By June it is high in the north or overhead in the late evening. In September it can still be seen earlier in the evening, looking west.

**1 March, 2am**
Looking east

**1 June, midnight**
Looking high in north

**1 September, 10pm**
Looking west

# A Quick Guide to **the Zodiac**

Although it's not obvious because of the Sun's glare, as Earth orbits the Sun, the position of the Sun gradually moves day by day against the background of stars on the celestial sphere. The path traced out by the Sun is called the ecliptic, and a band of the celestial sphere on either side of this path is known as the zodiac.

## Movements through the zodiac

As seen from Earth, all of the planets also gradually move, at varying speeds, against the starry background. Unlike the Sun, they do not strictly follow the line of the ecliptic, but they always stay within the band of the zodiac. Their movements are predominantly in the same direction as the Sun. In these movements, both Sun and planets pass through 13 constellations, known as zodiacal constellations.

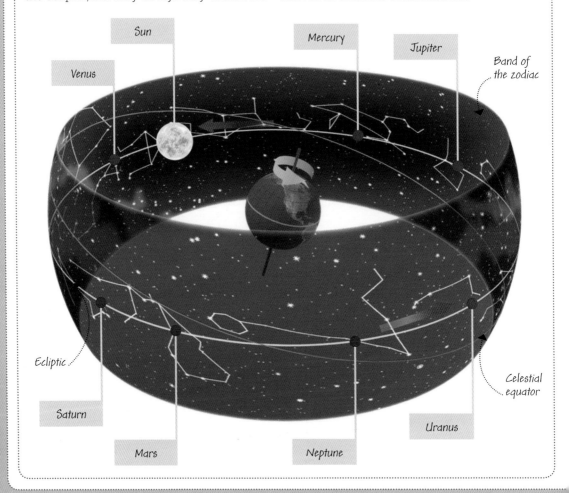

# THE ZODIAC

## Zodiac signs

Long ago astrologers adopted 12 of the 13 zodiacal constellations as "signs of the zodiac". For obscure reasons the thirteenth – Ophiuchus – was largely ignored by astrologers.

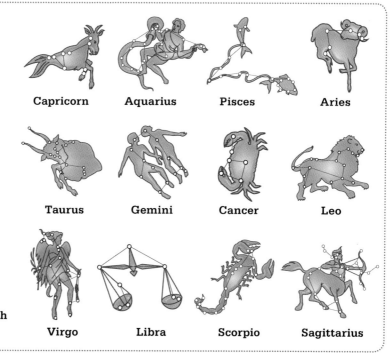

Capricorn · Aquarius · Pisces · Aries
Taurus · Gemini · Cancer · Leo
Virgo · Libra · Scorpio · Sagittarius

**Ophiuchus, the thirteenth zodiacal constellation**

## Retrograde motion

The planets predominantly travel through the zodiac in the same direction as the Sun, but sometimes for a few weeks or months one appears to move in the opposite direction. This is "retrograde motion", and mostly occurs when Earth overtakes another planet "on the inside" as they orbit the Sun. For example, Earth orbits faster than Mars and "overtakes" it every 25 months. As this happens, against the background of stars Mars seems to switch its direction of travel but resumes its normal motion once Earth has swung by.

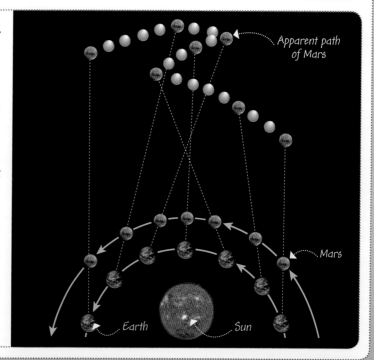

OBSERVATION TECHNIQUES

# Observing **Jupiter**

Jupiter is one of the easiest planets to spot with the naked eye: when present in a clear sky it shines brighter than any star, with a steady white light. It is closest to Earth – and at its brightest – during "opposition".

## Tracking Jupiter

The easiest way to spot Jupiter is to look for it in the night sky around the time of opposition. The chart below shows the months up to 2030 when the planet is at – or close to – opposition. The constellation that Jupiter is "in" at the time of opposition is also given – so in March 2016 Jupiter is situated just below the constellation figure of Leo.

To locate Jupiter around opposition, look into the southern sky (from the northern hemisphere) or overhead at around midnight; if you are in the southern hemisphere, look into the northern sky. Jupiter will be the brightest object in view, other than the Moon.

> **Remember** Jupiter only moves between constellations about once a year. When it's not in opposition, look for the nearest constellation figure and you may be able to spot Jupiter too.

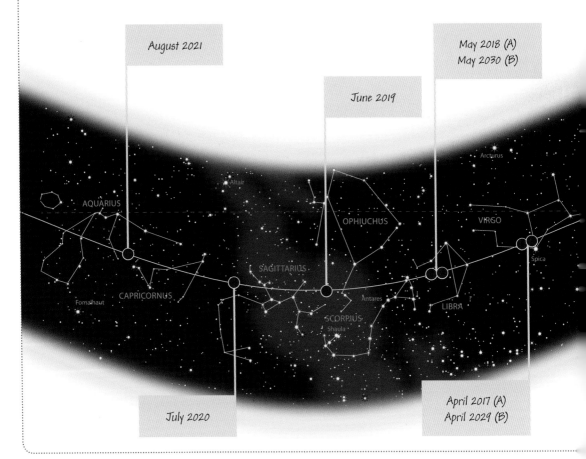

68

# JUPITER

## The view from Earth

Jupiter takes 12 years to orbit the Sun. As Earth takes just one year, it overtakes Jupiter "on the inside" every 13 months: this is known as "opposition", and is the best time to see the planet. Jupiter cannot be seen when it is in "conjunction" on the other side of the Sun, which also happens every 13 months.

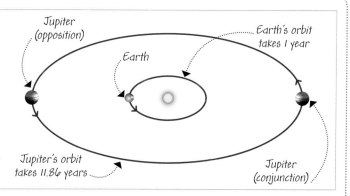

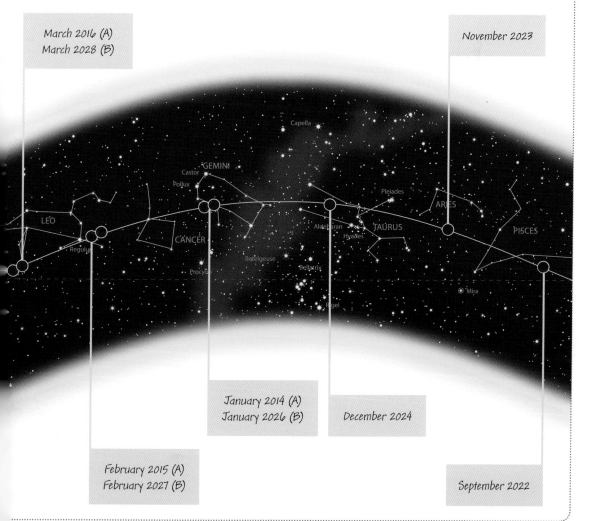

March 2016 (A)
March 2028 (B)

November 2023

January 2014 (A)
January 2026 (B)

December 2024

February 2015 (A)
February 2027 (B)

September 2022

# A Quick Guide to **Jupiter**

Easily the largest planet in the Solar System, Jupiter's mass is almost 2.5 times that of the other planets combined. It has a turbulent atmosphere with winds up to 625 kph (388 mph), and the strongest magnetic field of all the planets. Thanks to immense gravity, Jupiter has amassed a huge collection of moons.

## Orbit

Jupiter's orbit is highly elliptical – its distance from the Sun varies by 76.1 million km (47.2 million miles). It tilts only slightly as it spins; as neither hemisphere ever points markedly towards or away from the Sun, the planet has no seasons. Jupiter has a particularly fast spin, which causes a distinct bulge at its equator.

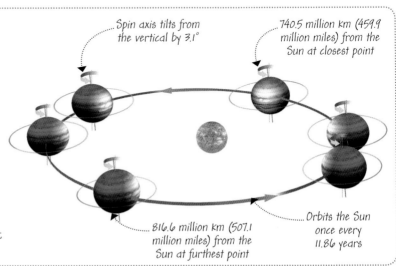

Spin axis tilts from the vertical by 3.1°

740.5 million km (459.9 million miles) from the Sun at closest point

816.6 million km (507.1 million miles) from the Sun at furthest point

Orbits the Sun once every 11.86 years

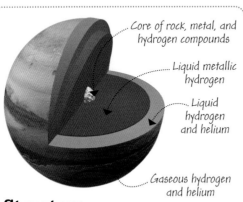

Core of rock, metal, and hydrogen compounds

Liquid metallic hydrogen

Liquid hydrogen and helium

Gaseous hydrogen and helium

## Structure

Jupiter's centre is a dense core thought to consist partly of rock. Surrounding this are liquid layers, mainly of hydrogen and helium, which gradually merge into an atmosphere of hydrogen and helium gas.

## The Great Red Spot

The largest and most obvious single feature on Jupiter's surface is the Great Red Spot, a vast persistent storm that has been raging for at least 300 years. The whole spot rotates anticlockwise about once every six days.

# JUPITER

## Jupiter profile

Jupiter is the fifth planet out from the Sun, and with a diameter of 142,984km (88,846 miles) it is the largest planet in the Solar System.

**Average distance from the Sun**
778.3 million km (483.6 million miles)

**Orbits the Sun once every**
11.86 years

**Rotates once every**
9 hours 55 minutes 30 seconds

**Greatest apparent magnitude**
-2.9 (see p.51)

**Furthest from Earth**
968.1 million km (601.5 million miles)

**Nearest to Earth**
588.5 million km (365.7 million miles)

**Greatest angular size in sky**
50.1 arcseconds (see p.12)

**Number of moons**
67 confirmed as of July 2013

**Size comparison**

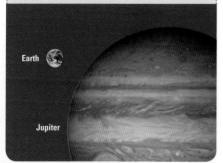

Earth

Jupiter

Distinctively banded atmosphere, affected by fierce storms and winds

The Great Red Spot is about 30,000km (18,600 miles) wide

Cloud-top temperature of -110°C (-160°F)

# KEY CONSTELLATIONS

## Observing **Perseus**

A prominent northern constellation, Perseus represents the mythological hero who slayed the gorgon Medusa. If you're in the northern hemisphere, this is one of the first constellations to identify and take a look at – the star Algol is a particularly interesting feature to observe as it periodically fluctuates in brightness.

### Perseus, the Greek hero

The figure of Perseus is depicted with his sword in one hand and Medusa's severed head in the other. The bright star Mirphak – on the hero's abdomen – lies at the centre of a cluster of fainter blue stars. Close to the sword handle is a double cluster of stars, which appear as a brighter patch in the Milky Way.

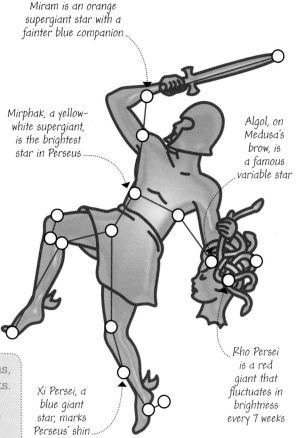

Miram is an orange supergiant star with a fainter blue companion

Mirphak, a yellow-white supergiant, is the brightest star in Perseus

Algol, on Medusa's brow, is a famous variable star

Xi Persei, a blue giant star, marks Perseus' shin

Rho Persei is a red giant that fluctuates in brightness every 7 weeks

**Remember** Once you've identified Perseus, look at the star Algol over successive nights. Every 69 hours it fades in brightness for about 10 hours. This is because it consists of two stars circling each other – it's an example of an eclipsing binary (see p.53).

### **Northern hemisphere** locators

For an observer at mid-latitudes (40°N), Perseus is visible at some point during the night for most of the year. The best seasons to view it are autumn and winter, when it is high in the south or overhead in the late evening.

**1 July, 4am**
Looking east

**1 November, midnight**
Looking high in south

**1 March, 8pm**
Looking west

# PERSEUS

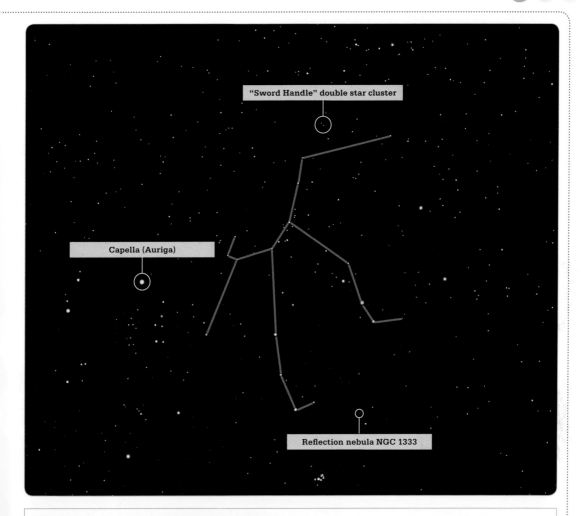

## Reflection nebula NGC 1333

Near Perseus' left foot is a cloud of dust that reflects light and other radiation from nearby stars. This "reflection nebula" is about 1,000 light years away. The red and green glows of infrared light were detected by NASA's Spitzer Telescope.

OBSERVATION TECHNIQUES

# Observing a **Meteor Shower**

Meteors are the name astronomers give to shooting stars – small grains of space dust that hit Earth's atmosphere at enormous speed and then burn up, producing streaks of light that shoot across the sky. At certain times of year, "showers" of meteors – many flashes, all emanating from the same point in the sky, called the radiant – can be observed.

## When and where to observe a shower

First identify the next meteor shower using the table below, bearing in mind the time of year and whether you're observing from the northern or southern hemisphere. Choose a clear night on, or close to, the peak date – preferably one without a full or waning gibbous Moon (see p.30) – and find a spot well away from light pollution with a good view of the full sky from which to observe. Start watching for meteors around midnight or in the early hours of the morning. By then the radiant for most showers will be fairly high above the horizon.

### Observation technique

First look into the sky and try to identify the constellation that contains the radiant – the point from which most of the meteors will emanate. The table below lists the relevant constellation for each shower. If necessary use a planisphere (see p.58) to help locate it in the sky. Give your eyes 20 minutes to adapt to the dark. Then sit or lie back, look up, and be patient. The flashes can happen anywhere in the sky and are usually quick and unpredictable – but over an hour or so you should see plenty.

### Diary dates: major annual meteor showers

| Date range | Peak dates | Constellation | Shower | Max per hour |
| --- | --- | --- | --- | --- |
| 28 Dec–12 Jan | 3–4 Jan | Boötes (N) | Quadrantids | 100 |
| 16–26 Apr | 22 Apr | Lyra (N) | Lyrids | 20 |
| 21 Apr–24 May | 6 May | Aquarius (S) | Eta Aquariids | 35 |
| 14 Jul–18 Aug | 28–29 Jul | Aquarius (S) | Southern Delta Aquariids | 20 |
| 15 Jul–25 Aug | 1 Aug | Capricornus (N, S) | Alpha Capricornids | 5 |
| 23 Jul–22 Aug | 12 Aug | Perseus (N) | Perseids | 80 |

# METEOR SHOWERS

**Perseid meteors and the Milky Way**

| Date range | Peak dates | Constellation | Shower | Max per hour |
| --- | --- | --- | --- | --- |
| 27 Sep–2 Dec | 30 Oct–7 Nov | Taurus (N, S) | Taurids | 10 |
| 5–30 Oct | 20 Oct | Orion (N, S) | Orionids | 35 |
| 14–21 Nov | 17 Nov | Leo (N, S) | Leonids | Variable |
| 6–18 Dec | 14 Dec | Gemini (N) | Geminids | 100 |
| 17–25 Dec | 22 Dec | Ursa Minor (N) | Ursids | 102 |

(N) Best viewed from northern hemisphere   (S) Best viewed from southern hemisphere   (N, S) Viewable from either hemisphere

# A Quick Guide to **Meteors and Meteorites**

Moving around in space within the Solar System are large amounts of dust and rock that have come from comets and fragmented asteroids. When a small grain of this material hits Earth's atmosphere and burns up, it is called a meteor; but if a large chunk penetrates the atmosphere and reaches the ground without vaporizing, it is termed a meteorite.

## Meteor showers

Meteors, or shooting stars, can be seen on any night, but showers tend to happen at certain times of year when Earth, in its orbit round the Sun, passes through a trail of dust left by a comet. For example, the Perseid meteor shower – which peaks in August – occurs when Earth passes into a dust trail left by comet Swift-Tuttle. The dust trail has a specific orientation in space, which means that all the shooting stars seem to originate from the same point in the sky, called the radiant.

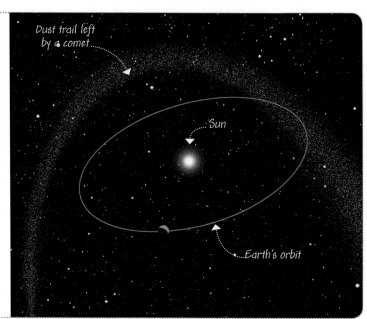

## Meteorites

Meteorites are classified according to their composition. Iron meteorites consist mainly of an iron-nickel alloy, stony meteorites are rocky, and stony-iron ones are a mixture of rock and iron-nickel alloy. Extremely rarely, a meteorite injures a person when it lands. On even rarer occasions, a large meteorite may create a massive impact crater.

**Iron meteorite**

**Stony meteorite**

**Stony-iron meteorite**

# METEORS AND METEORITES

Geminid Meteor over Iran

# KEY CONSTELLATIONS

## Observing **Cygnus**

During summer in the northern hemisphere, a beautiful constellation to look out for is that of Cygnus, the Swan. Its brightest star, Deneb, also belongs to a magnificent trio of stars known as the Summer Triangle. If you can spot this triangle, it will help you to locate Cygnus as well.

## Cygnus, the Swan

The constellation figure depicts a swan flying southward along the Milky Way – on a dark night you should be able to see this in the background. Deneb sits in the swan's tail. The other two stars of the Summer Triangle are in nearby constellations: they are Vega (Lyra), and Altair in Aquila.

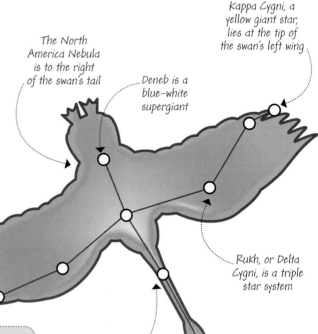

The North America Nebula is to the right of the swan's tail

Deneb is a blue-white supergiant

Kappa Cygni, a yellow giant star, lies at the tip of the swan's left wing

Rukh, or Delta Cygni, is a triple star system

Eta Cygni is close to an object that gives off X-rays: it may be the site of a black hole

Albireo is a double star consisting of a golden giant and a smaller blue companion

**Remember** Deneb may be 200,000 times more luminous than the Sun, but from Earth it looks fainter than the other two stars in the Summer Triangle. This is because it is more than 1,500 light years away compared to 25 and 17 light years for Vega and Altair.

## **Northern hemisphere** locators

From mid-latitudes (around 40°N), Cygnus is visible at some time during the night all year round. In summer it is almost overhead around midnight, and from autumn to mid-winter you should see it in the western sky earlier in the evening.

**1 May, 2am**
Looking east

**1 August, midnight**
Looking high in south

**1 November, 10pm**
Looking west

# CYGNUS

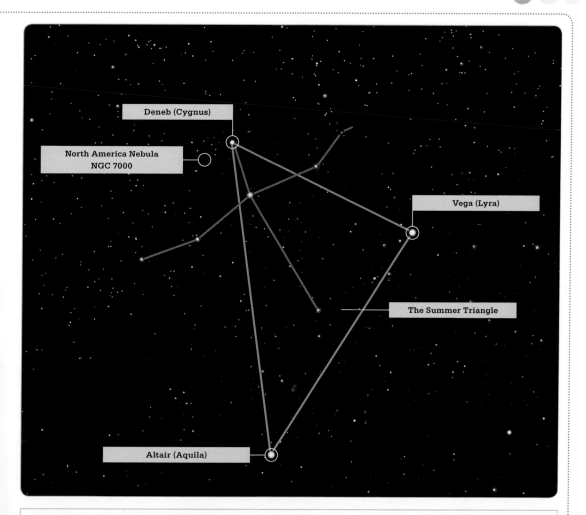

## North America Nebula NGC 7000

Close to Deneb is a slightly brighter patch of the Milky Way. Known as the North America Nebula after its resemblance to the continent's shape, this is an emission nebula (see p.116). You may need to use binoculars to see it clearly.

KEY CONSTELLATIONS

## Starhopping from the Summer Triangle

The brightest three stars in northern summer skies – Deneb, Vega, and Altair – form a distinct triangular signpost that can be used to identify nearby constellations. Try these "star hops":

**1 To Hercules** From Deneb, the faintest star in the Triangle, follow an imaginary line along the shortest side of the Triangle to Vega. Continue for about the same distance again, and you will come to the constellation of Hercules, the Strong Man. Look out for a distinctive four-star asterism within Hercules – the Keystone – just off to the side of your imaginary line.

**2 To Nunki in Sagittarius** Starting from Deneb, trace a line through Altair, the third star in the Summer Triangle. Continuing this line about the same distance again, and deviating a little to the left, you will arrive at a star called Nunki, which is the second brightest star in the constellation of Sagittarius.

**3 To the head of Draco** Deneb lies on the "tail" of Cygnus, the Swan. Using the chart (right) to help, find the central star of Cygnus, called Sadr. From here visualize a line to the star Rukh, on the wing of the Swan. Extend the line on and you will come to two bright stars in the constellation of Draco, the Dragon. Next to these are two fainter stars; together the four stars make up the head of the Dragon.

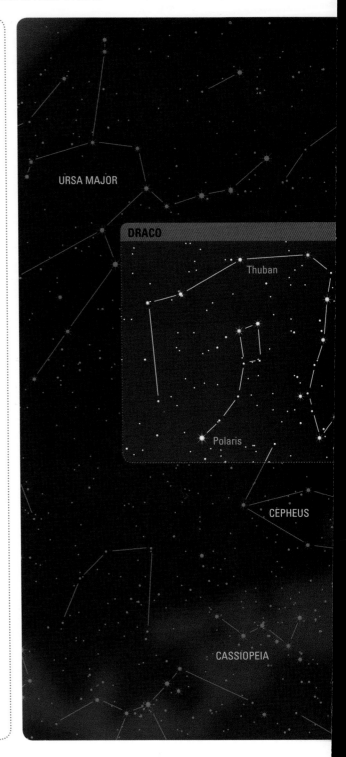

# STARHOPPING FROM THE SUMMER TRIANGLE

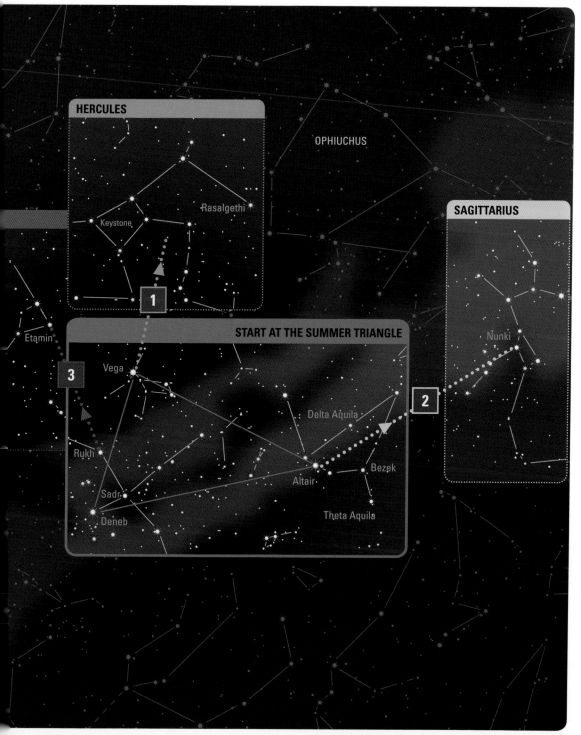

OBSERVATION TECHNIQUES

# Observing **Venus**

Venus, our nearest neighbour, is unmistakable and spectacular when present in the evening or early morning sky. It outshines everything except for the Sun and Moon.

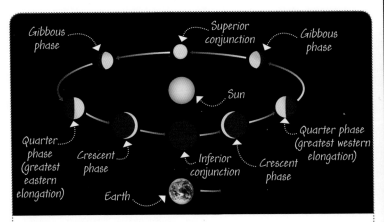

## The view from Earth

Venus orbits the Sun faster than Earth, overtaking it "on the inside" every 19 months. We don't see it at the point of each overtake (inferior conjunction) because it sits between Earth and the Sun, but afterwards, for about 6 months, Venus may be seen in the east at dawn. It then moves out of view on the other side of the Sun (superior conjunction) for a few months. Venus next appears in the west around and after sunset for 6 months until it overtakes Earth again.

## Locating Venus

To view Venus, choose a date within 6 weeks of "greatest elongation" – Venus will be at its maximum distance from the Sun, and easier to see. This is not necessarily when the planet is at its brightest; for evening apparitions this follows a few weeks later. You need binoculars or a telescope to see the Moon-like phases Venus displays over each 19-month cycle.

**LOOK EAST in the early morning, before and around dawn**

| Date | Brightest | Greatest Elongation |
|---|---|---|
| Jan–Aug 2014 | 3–24 Feb | 22 Mar |
| Sep 2015–Mar 2016 | 10 Sep–6 Oct | 26 Oct |
| Apr–Oct 2017 | 18 Apr–7 May | 3 Jun |
| Nov 2018–May 2019 | 20 Nov–14 Dec | 6 Jan |
| Jun 2020–Jan 2021 | 26 Jun–28 Jul | 13 Aug |
| Feb–Aug 2022 | 31 Jan–22 Feb | 20 Mar |
| Sep 2023–Mar 2024 | 8 Sep–2 Oct | 23 Oct |
| Apr–Oct 2025 | 15 Apr–5 May | 1 Jun |

# VENUS

Venus at sunset over Selsey Beach in Sussex

**LOOK WEST in the evening, around and after sunset**

| Date | Greatest Elongation | Brightest |
| --- | --- | --- |
| Feb–Aug 2015 | 6 Jun | 22 Jun–24 Jul |
| Sep 2016–Mar 2017 | 12 Jan | 7–27 Feb |
| Apr–Oct 2018 | 17 Aug | 9 Sep–5 Oct |
| Nov 2019–May 2020 | 24 Mar | 18 Apr–5 May |
| Jun 2021–Jan 2022 | 29 Oct | 22 Nov–18 Dec |
| Feb–Jul 2023 | 4 Jun | 20 Jun–22 Jul |
| Aug 2024–Feb 2025 | 10 Jan | 4–24 Feb |
| Apr–Oct 2026 | 15 Aug | 8 Sep–3 Oct |

**Tip** If you look into the sky around or after sunset, or around dawn as appropriate on your chosen date, you will normally have no problem identifying Venus as by far the brightest object in that part of the sky. You are likely to see it again in roughly the same place on several subsequent evenings or mornings.

# QUICK GUIDE

## A Quick Guide to **Venus**

Venus is similar to Earth in size and structure, but differs greatly in other respects. An unbroken blanket of dense clouds permanently envelopes the planet, and underneath lies a gloomy, scorching hot, waterless and lifeless surface. The volcanic activity that shaped Venus is thought to have occurred some 500 million years ago.

### Orbit

Venus has the most nearly circular orbit of all the planets. It spins in the opposite direction to the other planets – except for Uranus, which spins on its side. The rotation period of over 243 Earth days is longer than that of any other planet, so a day on Venus lasts longer than the Venusian year (around 225 Earth days).

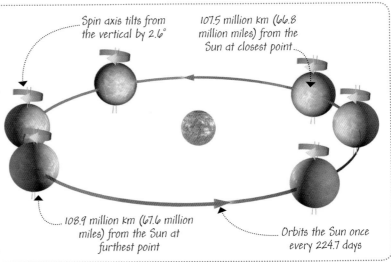

*Spin axis tilts from the vertical by 2.6°*

*107.5 million km (66.8 million miles) from the Sun at closest point*

*108.9 million km (67.6 million miles) from the Sun at furthest point*

*Orbits the Sun once every 224.7 days*

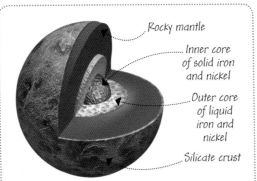

*Rocky mantle*
*Inner core of solid iron and nickel*
*Outer core of liquid iron and nickel*
*Silicate crust*

### Structure

Venus has a solid metallic inner core, a molten metallic outer core, and a thick rocky mantle overlain by a thin crust. An extremely thick atmosphere – mainly carbon dioxide – surrounds the planet.

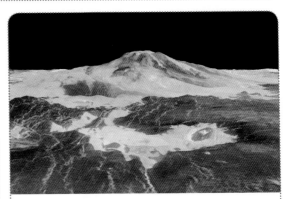

### Maat Mons

Images of Venus's surface sent back by probes reveal a barren landscape shaped by huge volcanoes. Maat Mons is the largest volcano on Venus, rising 5km (3 miles) above the surrounding plains.

## Venus profile

Venus is the second planet out from the Sun. With a diameter of 12,104km (7,521 miles) it is the third smallest planet in the Solar System.

**Average distance from the Sun**
108.2 million km (67.2 million miles)

**Orbits the Sun once every**
224.7 days

**Rotates once every**
243 days 16 hours 28 minutes

**Greatest apparent magnitude**
-4.6 (see p.51)

**Furthest from Earth**
261 million km (162 million miles)

**Nearest to Earth**
38.2 million km (23.7 million miles)

**Greatest angular size in sky**
66 arcseconds (see p.12)

**Number of moons**
0

**Size comparison**

*Hostile, desolate surface with no liquid water*

*Vast ancient lava flows*

*Average surface temperature of 464°C (867°F)*

OBSERVATION TECHNIQUES

# Observing **Artificial Satellites**

Far above Earth's surface are thousands of objects that have been put there by people. Known as artificial satellites, they range from simple laser-reflecting spheres to complex space telescopes and a massive manned spacecraft, the International Space Station (ISS). It's easier to observe some of these objects from Earth than you might think.

## Satellite orbits

Satellites follow a variety of different orbits around Earth. These orbits are established at launch and continuously fine-tuned by signals sent from control stations on Earth.

Satellites in low Earth orbit – most notably the International Space Station and Hubble Space Telescope (HST) – are most easily seen from the ground as pinpricks of light, caused by the reflection of sunlight, moving steadily across the sky. Different low Earth satellites cross the sky in various directions – from east to west, north to south, or obliquely.

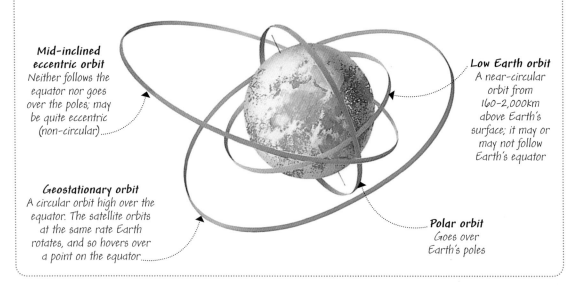

**Mid-inclined eccentric orbit**
Neither follows the equator nor goes over the poles; may be quite eccentric (non-circular)

**Low Earth orbit**
A near-circular orbit from 160–2,000km above Earth's surface; it may or may not follow Earth's equator

**Geostationary orbit**
A circular orbit high over the equator. The satellite orbits at the same rate Earth rotates, and so hovers over a point on the equator

**Polar orbit**
Goes over Earth's poles

## Iridium flares

Many artificial satellites cross the sky each night, but one group is responsible for the spectacular phenomena known as "Iridium flares" – dazzling flashes of light lasting up to 10 seconds. Flares occur when the antennae of Iridium communications satellites reflect an intense beam of sunlight towards Earth; they are predictable, and internet calculators will tell you when to expect one in your area.

ARTIFICIAL SATELLITES

## International Space Station

The ISS is the most noticeable of all satellites – a sharp, very bright star-like dot that moves across the sky at a slow, steady pace. Typical passes last 1–7 minutes: the ISS may traverse a large area of sky before disappearing into Earth's shadow, or over the horizon.

For a particular observing location there tend to be regular "flyby seasons" for the ISS that last for about a week, during which the space station may make several visible passes, usually soon after sunset, or before sunrise, when it is basking in sunlight. NASA offers a web-based tool listing when the ISS will be visible from your area, as well as the location in the sky where it will first appear and where it will move to before disappearing. A similar tool exists listing times to view the Hubble Space Telescope (HST).

# The science of **Gravity and Orbits**

Gravity is a force that attracts objects together – its strength depends on the masses of the bodies and distance between them. The gravitational attraction between everyday-sized objects, such as billiard balls placed next to each other, is tiny. But if at least one object is big – like a planet – or the overall amount of matter is large, gravity has a profound effect.

## What does gravity do?

Some important effects of gravity are shown below. Gravity holds large objects, such as stars, together, and keeps various bodies in orbit around others. Because gravity causes matter to aggregate, without it galaxies, stars, and planets would never have formed. Instead, the Universe would consist of just a thin – though vast – cloud of gas.

Holds stars together: without it, a star would disperse into space

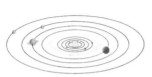
Keeps planets in orbit around the Sun

Keeps everything in the Milky Way galaxy in orbit around the galactic centre

Keeps the Moon in orbit around Earth

## Newton's idea about gravity

The scientist Isaac Newton was the first person to realise that the same force that causes an apple to fall to the ground on Earth – gravity – also causes the Moon to orbit Earth, Earth and other planets to orbit the Sun, and so on. Newton reasoned that an object remains stationary or moves at a constant speed in a straight line unless a force acts upon it. In the case of the Moon, the force of gravity continuously acts to pull the Moon towards the Earth. This pull changes what would otherwise be the Moon's straight-line motion into an orbit, as shown in the diagram (right).

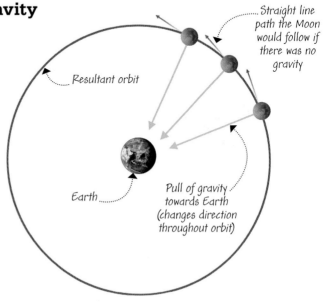

# GRAVITY AND ORBITS

## Common centre of gravity

We usually think of a small object orbiting a much larger one, but in space there are many examples of two bodies – typically stars – of more equal size that orbit around each other.

More precisely, the two bodies each orbit the overall centre of gravity, sometimes called the common centre of mass, of the whole system, as shown in the diagram below.

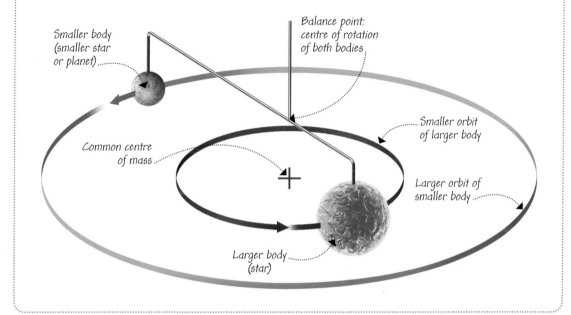

## Orbit shapes and eccentricity

All naturally occurring orbits have the shapes of ellipses – "stretched" circles. The degree to which an elliptical orbit differs from a perfect circle is called its "eccentricity". Orbits that are quite close to being circles – such as the Moon's orbit around Earth – are said to be of low eccentricity, while orbits that are very elongated ellipses are of high eccentricity. The orbits of comets around the Sun, for example, vary from highly to exceedingly eccentric.

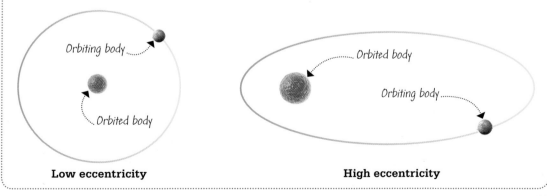

# Observing a **Solar Eclipse**

A solar eclipse occurs when the Moon blocks sunlight from reaching parts of Earth's surface. There are two or three partial eclipses each year, so you'll have a chance to see one every few years. A total eclipse – where the Sun's light is completely blocked out by the Moon and for a few minutes day turns to night – is much rarer. If you want to see one, you may need to travel!

## Appearance of an eclipse

In a partial or annular eclipse, the Moon obscures a gradually enlarging, and then gradually diminishing, portion of the Sun. The process lasts anything from a few minutes up to an hour and a half. As our view of the Sun is blocked it becomes darker; and as the obscured area diminishes, light levels return to normal.

A total eclipse is more spectacular. The Sun's disc is completely blocked – typically for 1–3 minutes but sometimes for 7 minutes or more. In this period of totality it becomes as dark as night, stars appear, temperatures drop, and the Sun's corona, a hot, irregularly shaped part of its atmosphere, surrounds a black disc in the sky.

# SOLAR ECLIPSE

## Viewing an eclipse

Never look at the Sun directly with the naked eye or through sunglasses, binoculars, a telescope, or a camera: its intense light can cause eye damage. Use purpose-designed eclipse-viewing spectacles to view an eclipse before and after totality.

## How a solar eclipse works

Eclipses are either partial, total, or annular (when the Moon blocks out all but a narrow ring of the Sun's disc). Annular eclipses occur when the Moon is further from Earth than average, and therefore looks smaller than normal. In a partial eclipse, viewers across a broad swathe of Earth's surface see the Sun partly hidden for a short time.

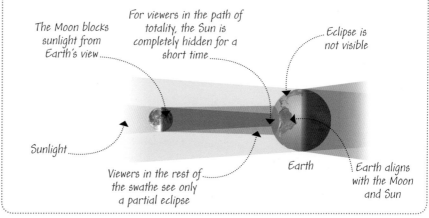

- The Moon blocks sunlight from Earth's view
- For viewers in the path of totality, the Sun is completely hidden for a short time
- Eclipse is not visible
- Sunlight
- Viewers in the rest of the swathe see only a partial eclipse
- Earth
- Earth aligns with the Moon and Sun

## Diary dates: total solar eclipses

The table shows when and where you might see a total eclipse, depending on cloud coverage; at the same time a wider area will experience a partial eclipse.

| Date | Location |
| --- | --- |
| 20 Mar 2015 | North Atlantic, Svalbard, Arctic Ocean |
| 9 Mar 2016 | Indonesia, Western and Central Pacific Ocean |
| 21 Aug 2017 | North Eastern Pacific Ocean, United States, Central Atlantic Ocean |
| 2 Jul 2019 | South Pacific Ocean, Chile, Argentina |
| 14 Dec 2020 | Southeast Pacific Ocean, Chile, Argentina |
| 4 Dec 2021 | South Atlantic Ocean, Antarctica |
| 8 Apr 2024 | Central Pacific Ocean, Mexico, United States, Canada |
| 12 Aug 2026 | Greenland, North Atlantic Ocean, Spain |

**Remember**
It is only safe to observe the Sun with the unprotected eye during totality.

# 2
# Build On It

This chapter switches attention to some objects or classes of objects in the night sky that look great through binoculars. These include details on the Moon's surface, such as its seas and craters; a number of sparkling star clusters; some prominent and beautiful nebulae, including the Orion nebula; the Andromeda Galaxy; bright comets; and the planets Mars and Mercury. You'll learn about a few more constellations – mainly those that contain good objects for binocular viewing. And early in the chapter you'll also find a short guide to choosing and using binoculars.

**The Pleiades star cluster in the constellation of Taurus**

TOOLS AND EQUIPMENT

# Choosing and using **binoculars**

Binoculars are ideal for taking visual tours across the night sky. If you own a pair for watching wildlife you can probably use them to observe a wide range of astronomical objects, but if you're buying binoculars specifically for astronomy, there are a few things to look out for.

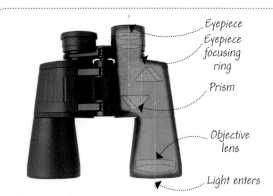

**Standard porro-prism binoculars**

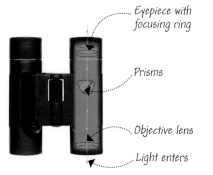

**Compact roof-prism binoculars**

## Types of binoculars

Binoculars magnify objects and – because they collect more light than the naked eye – make dim objects look brighter. There are two main designs. The porro-prism design has the classic dog-leg shape; they can be wide and bulky but the large aperture of the objective (front) lens means that they tend to have better, brighter optics than the straighter, more compact roof-prism binocular design.

Binoculars are labelled with two numbers, for example 10 x 50. The first number is the magnifying power. High magnification might seem attractive, but can amplify every jolt as you try to hold them steady. A magnification of 10 is a good compromise. The second number is the diameter of the objective lens, and here, the bigger the better. For astronomy, look for an objective lens of at least 50mm.

## Using binoculars comfortably

Its important to be as comfortable as possible. Try reclining on a deckchair, lounger, or waterproof rug on the ground; this will allow you to view the sky at higher angles without straining your neck. Alternatively, sit with your elbows supported by your knees, or rest them on the top of a fence or wall to provide support and stability. Use a tripod for larger, heavier binoculars. Some (pricier) binoculars employ image stabilization technology to reduce the effects of hand shake.

Sitting and placing the elbows on the knees can support the weight of the binoculars and keep them steady

# Focusing

Before using your binoculars on the night sky, make sure they are focused on a distant object. The Moon is an obvious choice; if it's not present, try focusing on a cluster of stars. Most binoculars have two focus adjustment wheels – one on the central bar, which moves both eyepieces, and another on one of the eyepieces, which focuses just that eyepiece – the procedure below explains how to focus them. If each eyepiece has its own focusing ring, adjust each eyepiece in turn (closing the other eye), then look through both eyepieces.

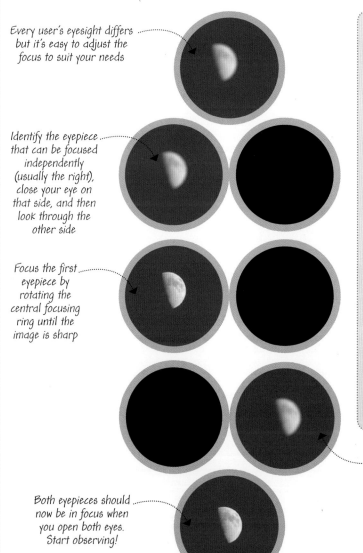

*Every user's eyesight differs but it's easy to adjust the focus to suit your needs*

*Identify the eyepiece that can be focused independently (usually the right), close your eye on that side, and then look through the other side*

*Focus the first eyepiece by rotating the central focusing ring until the image is sharp*

*With your other eye open (and the first one closed), adjust the focusing ring on the eyepiece until the image becomes sharp*

*Both eyepieces should now be in focus when you open both eyes. Start observing!*

## Messier numbers

In this chapter you will find that some of the sky objects referred to have an alphanumeric descriptor attached. For example, the Pleiades star cluster is also referred to as M45 and a particular galaxy in Centaurus is also referred to as NGC 5128. Descriptors starting with an "M" are called Messier numbers; they refer to the object's placement in a catalogue of 110 nebulous-looking sky objects compiled by the French astronomer, Charles Messier. A descriptor starting with "NGC" or "IC" indicates the object's placement in two larger catalogues: the New General Catalogue and the Index Catalogue.

OBSERVATION TECHNIQUES

# Observing **the Moon through binoculars**

Armed with a pair of binoculars, one of the most obvious targets to turn to is the Moon. Through binoculars, vague light and dark areas on the Moon's surface become more obvious and detailed, and you should be able to make out a number of the Moon's largest craters.

## Light and shadow

You don't have to wait for the Moon to be full to look at it through binoculars. It can be just as rewarding – in fact, often more interesting – to look at it during one of its crescent, half, or gibbous phases. You can see more detail when the Sun strikes irregularities such as craters at an angle. The dividing line between day and night on the Moon is called the "terminator".

### Remember
From the southern hemisphere the Moon appears to be upside down – compare the image on p.26. You see the same features, just rotated around 90–180° from their appearance here.

## Seas and craters

Using binoculars, try to identify some of the features on the Moon's near side – the face always turned to Earth. You'll see that half of it consists of rough-looking light areas: these are highland regions. The smooth, dark areas were believed by the ancient Greeks to be water so they named them maria or "seas"; in fact they are lowland basins blasted out by impacts during the Moon's early existence that later filled with flows of dark lava. Two of the largest are the Mare Serenitatis ("Sea of Serenity") and Mare Imbrium ("Sea of Rains").

Other features to look out for include large ray craters, like the Tycho crater, formed when impacting objects hit the surface and sent out splashes of molten rock. It's estimated that the Moon's near side is pitted with about 300,000 craters more than 1km (0.6 miles) wide.

**Mare Crisium**

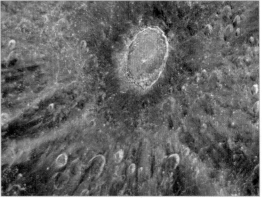

**Tycho crater**

# THE MOON

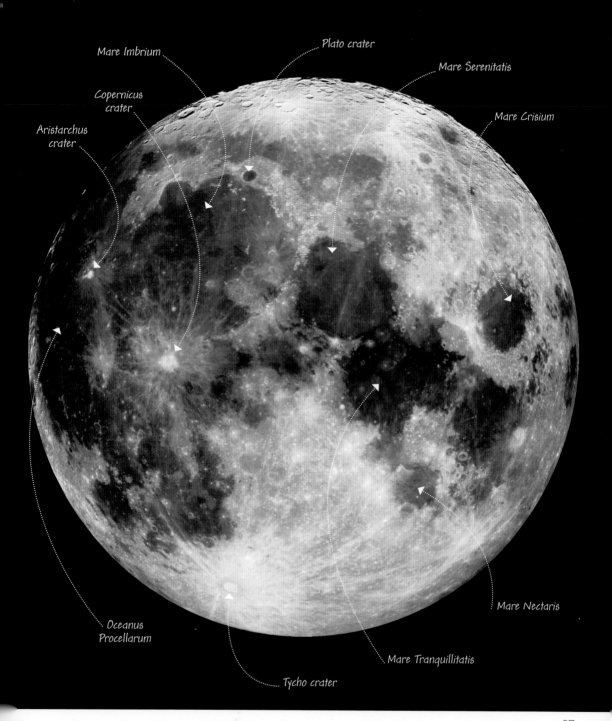

# QUICK GUIDE

## A Quick Guide to **the Moon**

The Moon is much smaller than Earth, with 26 per cent of Earth's diameter and only 1.2 per cent of its mass. However, when full, it is the brightest object in the night sky, and its gravity influences Earth's oceans. The Moon's surface is a dry, grey, lifeless place, covered in countless impact craters, with a negligible atmosphere.

## Orbit

The Moon has a slightly elliptical orbit – at its closest to Earth it is about 10 per cent closer than at its further point. It takes 27.32 days to spin on its axis, the same as the time it takes to orbit Earth. As a result one side of the Moon, its "near side", always faces Earth; no matter where you live, you'll see the same face.

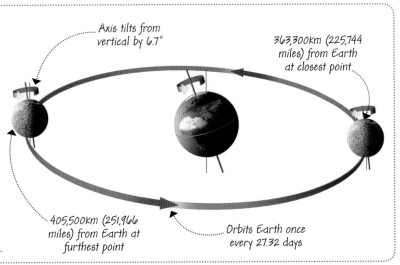

Axis tilts from vertical by 6.7°

363,300km (225,744 miles) from Earth at closest point

405,500km (251,966 miles) from Earth at furthest point

Orbits Earth once every 27.32 days

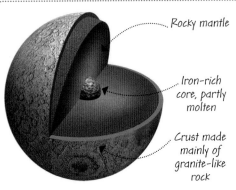

Rocky mantle

Iron-rich core, partly molten

Crust made mainly of granite-like rock

## Structure

Like Earth, the Moon has an iron-rich core, a mantle, and a crust, but the core is relatively small. The Moon's atmosphere is extremely thin, with a total mass of less than 25,000kg (24.6 tons).

## Human exploration

Twelve American astronauts (on six separate missions) visited the Moon between 1969 and 1972. They explored limited areas of its surface, took samples of Moon rocks, and left their footprints in the lunar soil.

# THE MOON

## Moon profile

*The Moon is Earth's only natural satellite. With a diameter of 3,475km (2,159 miles) it is the Solar System's 5th largest planetary satellite.*

**Average distance from Earth**
385,400km (239,477 miles)

**Orbits Earth once every**
27 days 7 hours 43 minutes

**Rotates once every**
27 days 7 hours 43 minutes

**Greatest apparent magnitude**
-12.9 (see p.51)

**Furthest from Earth**
405,500km (251,966 miles)

**Nearest to Earth**
363,300km (225,744 miles)

**Greatest angular size in sky**
34.1 arcminutes (see p.12)

**Age**
4.5 billion years

**Size comparison**

Heavily cratered surface

Surface temperature varies from up to 125°C (257°F) during the day down to -170°C (-274°F) at night

# KEY CONSTELLATIONS

## Observing **Taurus**

An ancient zodiacal constellation, Taurus has been recognized since Babylonian times and is visible from both hemispheres. Its attractions include two large open star clusters – the Hyades and Pleiades – which are excellent targets for binocular viewing.

## Taurus, the Bull

In the classic constellation figure, only the front half of a bull is represented. An impressively bright red star, Aldebaran, marks one eye; both the Hyades and more compact Pleiades are easy to find once you have identified it. The Taurid meteors, visible every November, appear to originate just south of the Pleiades.

**Remember** Taurus is most easily observed during November and December. It is difficult or even impossible to see in May and June because it is too close to the Sun.

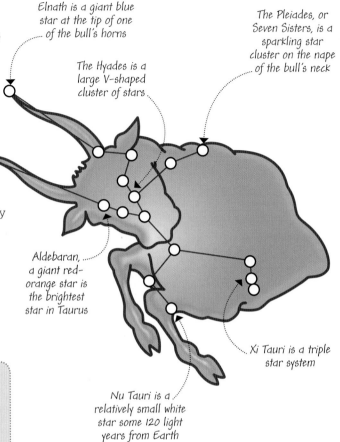

Elnath is a giant blue star at the tip of one of the bull's horns

The Hyades is a large V-shaped cluster of stars

The Pleiades, or Seven Sisters, is a sparkling star cluster on the nape of the bull's neck

Aldebaran, a giant red-orange star is the brightest star in Taurus

Nu Tauri is a relatively small white star some 120 light years from Earth

Xi Tauri is a triple star system

## **Northern hemisphere** locators

From mid-latitudes (around 40°N), Taurus starts to become visible from early August in the eastern sky – but only in the early morning hours. By December it is prominent and easily seen in the southern sky around midnight.

**1 August, 4am**
Looking east

**1 December, midnight**
Looking high in south

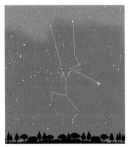

**1 April, 8pm**
Looking west

# TAURUS

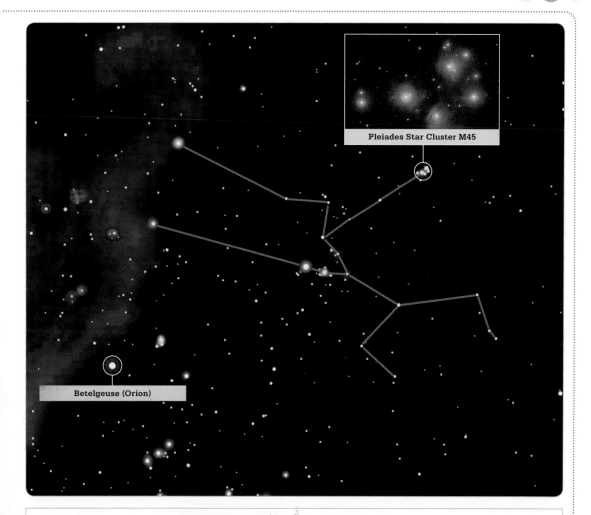

Pleiades Star Cluster M45

Betelgeuse (Orion)

## Southern hemisphere locators

From mid-latitudes (around 40°S) Taurus is visible low in the north – and upside down compared to the northern hemisphere – in November and December at around midnight. By February it is in the northwest earlier in the evening.

**1 October, 2am**
Looking northeast

**1 December, midnight**
Looking north

**1 February, 10pm**
Looking northwest

# A Quick Guide to **Star Clusters**

Large associations of stars that share a common origin are known as clusters. Two types have been identified in the Milky Way galaxy: "open clusters" are loose groupings of dozens or hundreds of relatively young stars, and "globular clusters" are tight-knit balls of up to a million stars, usually of ancient origin.

## Clusters in the Milky Way

Open clusters occur along the Milky Way's spiral arms, often near the nebulae in which they formed. Each star follows its own orbit around the galactic centre, so over millions of years the clusters tend to disperse. Globular clusters are permanently bound by the gravity of their close-packed stars. They orbit above and below the plane of the galactic disc.

A few globular clusters orbit close to the central bulge

Globular clusters move in long, elliptical orbits outside the galactic disc

Open clusters are found in the galaxy's spiral arms

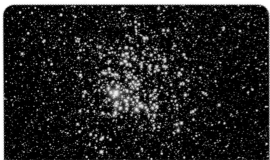

### Open star cluster
Irregular in shape, open clusters usually contain between 100 and 10,000 stars, all of which formed around the same time. Bright blue stars indicate a young cluster, while stars in older clusters are yellower.

### Globular cluster
Large, roughly spherical globular clusters are densely packed with many thousands of ancient red and yellow stars. They are among the oldest stars in the galaxy and help determine the age of the Universe.

# STAR CLUSTERS

Open star clusters M35 (upper left) and NGC 2158 (lower right)

# KEY CONSTELLATIONS

## Observing **Cassiopeia**

A distinctive constellation of the northern sky, Cassiopeia is located in a rich area of the Milky Way, full of star clusters that are a fine sight through binoculars. The large W-shape formed by its five main stars – or M-shape depending on your perspective – is one that is very easy to recognize.

**Remember** If you play your binoculars all around Cassiopeia, you'll come across many beautiful star clusters. When doing this you are looking in the direction of the Milky Way, which is where most of the gas and dust that creates star clusters resides.

### Cassiopeia, the Queen

According to Greek mythology, Cassiopeia was an Ethiopian Queen who – because of her vanity – was condemned by Poseidon to circle the north celestial pole forever. In her constellation figure Cassiopeia is classically depicted in a sitting position – and in this representation she is seen admiring herself in a mirror as she endlessly circles the northern sky. Four of the five stars that define Cassiopeia's figure are of roughly equal brightness; the fifth (known as Segin) is slightly fainter.

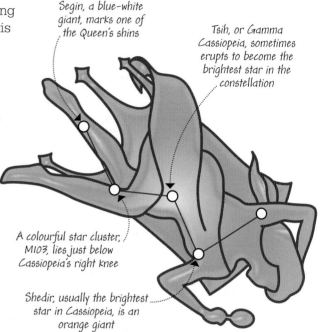

Segin, a blue-white giant, marks one of the Queen's shins

Tsih, or Gamma Cassiopeia, sometimes erupts to become the brightest star in the constellation

A colourful star cluster, M103, lies just below Cassiopeia's right knee

Shedir, usually the brightest star in Cassiopeia, is an orange giant

### **Northern hemisphere** locators

For an observer at mid-latitudes (around 40°N) Cassiopeia is visible in the northern sky for most of the night every night – but the constellation does sometimes partly dip below the horizon, for example at around 2am in March.

**1 June, 3am**
Looking northeast

**1 October, midnight**
Looking high in north

**1 February, 9pm**
Looking northwest

# CASSIOPEIA

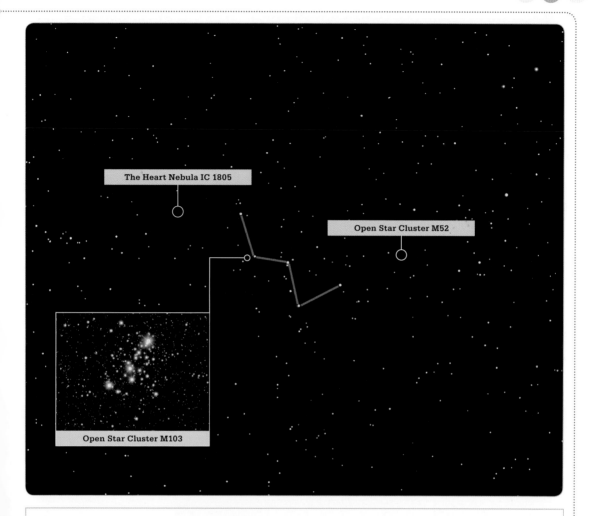

## The Heart Nebula IC 1805

This wispy star-forming cloud of dust and gas within the Milky Way galaxy can be found below Cassiopeia's legs. The nebula is named for its roughly heart-shaped outline, and is about 7,500 light years away from Earth.

# KEY CONSTELLATIONS

## Observing **Hercules**

The large constellation of Hercules represents the hero of Greek myth who undertook 12 labours. Within it you can find a magnificent star cluster known as the Hercules Globular Cluster. This is a fine object to examine through binoculars.

## Hercules, the Strong Man

Although it is most easily viewed from the northern hemisphere, parts or all of Hercules may also be seen from southern hemisphere locations. The figure is arranged around the Keystone – a quadrilateral asterism – and shows Hercules with a club in one hand and the multiple severed heads of Cerberus, a monstrous dog, in the other. Look for the Keystone to locate the constellation.

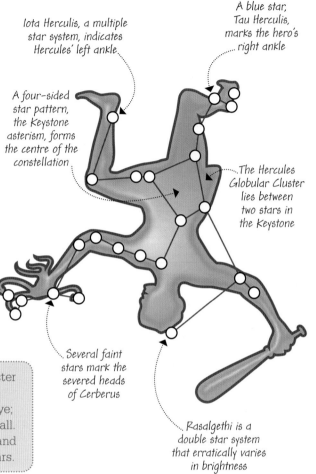

Iota Herculis, a multiple star system, indicates Hercules' left ankle.

A blue star, Tau Herculis, marks the hero's right ankle.

A four-sided star pattern, the Keystone asterism, forms the centre of the constellation.

The Hercules Globular Cluster lies between two stars in the Keystone.

Several faint stars mark the severed heads of Cerberus.

Rasalgethi is a double star system that erratically varies in brightness.

**Remember** The Hercules Globular Cluster (M13), situated between two stars in the Keystone, can be seen with the naked eye; through binoculars it looks like a fuzzy ball. It is 25,000 light years away from Earth and contains over 300,000 closely packed stars.

## **Northern hemisphere** locators

From mid-latitudes Hercules may be seen during late winter in the east, but only in the early morning hours. By June it is high in the south or overhead at midnight. On autumn evenings Hercules is visible in the western sky.

**1 February, 3am**
Looking east

**1 June, midnight**
Looking high in south

**1 October, 9pm**
Looking west

# HERCULES

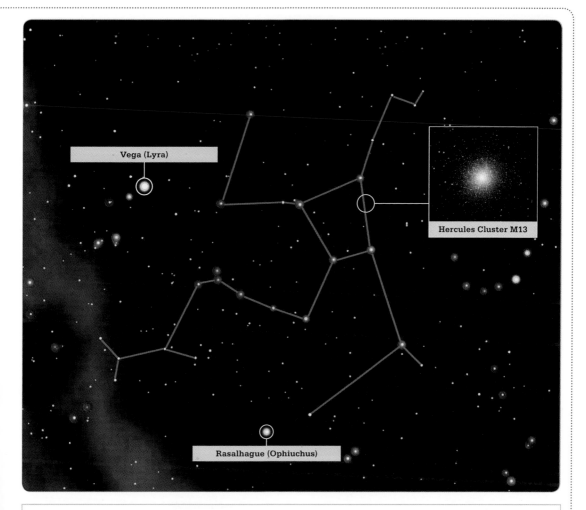

Vega (Lyra)

Hercules Cluster M13

Rasalhague (Ophiuchus)

## Southern hemisphere locators

From mid-latitudes, most of Hercules can be seen above the northern horizon around midnight during June. It appears in the same place during July evenings, or in the early morning hours throughout May.

**1 May, 4am**
Looking north

**1 June, midnight**
Looking north

**1 July, 8pm**
Looking north

KEY CONSTELLATIONS

# Observing **Centaurus**

This large constellation contains several objects of interest, including our nearest star neighbours and a massive globular star cluster, Omega Centauri – an excellent target for binoculars. Unfortunately Centaurus is blocked off from the sight of most northern hemisphere skywatchers.

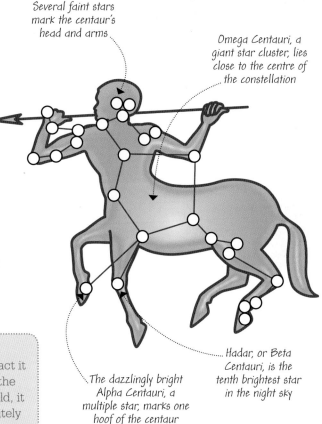

Several faint stars mark the centaur's head and arms

Omega Centauri, a giant star cluster, lies close to the centre of the constellation

The dazzlingly bright Alpha Centauri, a multiple star, marks one hoof of the centaur

Hadar, or Beta Centauri, is the tenth brightest star in the night sky

## Centaurus, the Centaur

The figure depicted is a mythical being with his feet embedded in the Milky Way. Its brightest star, known as Alpha Centauri or Rigil Kentaurus, is a system of three gravitationally bound stars. Other than the Sun, these are the closest stars to Earth – around 4.2 light years away. The faintest of the three, Proxima Centauri, is individually the closest.

**Remember** To the naked eye Omega Centauri looks like a smudged star. In fact it is a globular star cluster, the largest in the Milky Way. More than 12 billion years old, it holds over a million stars – and is definitely worth admiring through binoculars.

## **Centaurus A Galaxy** NGC 5128

This unusual galaxy in the centaur's back is just visible through binoculars. The dark band of dust that cuts across is thought to result from a merger of two earlier galaxies. It is 13 million light years away, with a massive black hole at its centre.

# CENTAURUS

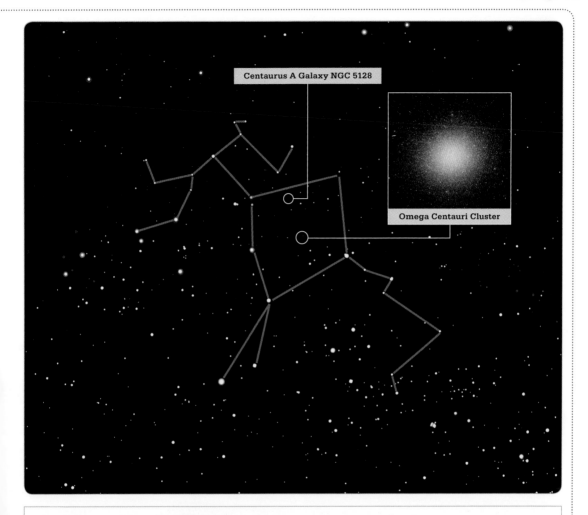

## Southern hemisphere locators

From mid-latitudes Centaurus is visible for part of the night in most months of the year. Optimal viewing times are March to May, when it is nearly overhead at midnight, or from June to August in the south-western evening sky.

**1 December, 3am**
Looking southeast

**1 April, midnight**
Looking high in south

**1 August, 9pm**
Looking southwest

# Observing **Mars**

Mars is fairly easy to see. It varies considerably in brightness, but it is always visible to the naked eye when present, shining with a reddish light. Through binoculars it appears as a distinct disc.

## Tracking Mars

The easiest way to spot Mars is to look for it at around the time of opposition. The chart below shows the months up to 2030 when Mars is at, or close to, opposition. For each month, the constellation that Mars is "in" at the time is also given – so in May 2016 Mars is in Scorpius, between the scorpion's claws.

To locate Mars during opposition, just look into the southern sky (from the northern hemisphere) or overhead at about midnight; if you are in the southern hemisphere, look north. You should see Mars as one of the brightest objects in view.

**Tip** To spot Mars at times other than opposition, find out which constellation it is in, then use a planisphere or "app" to work out when and where to see that constellation. This should help you to spot Mars too.

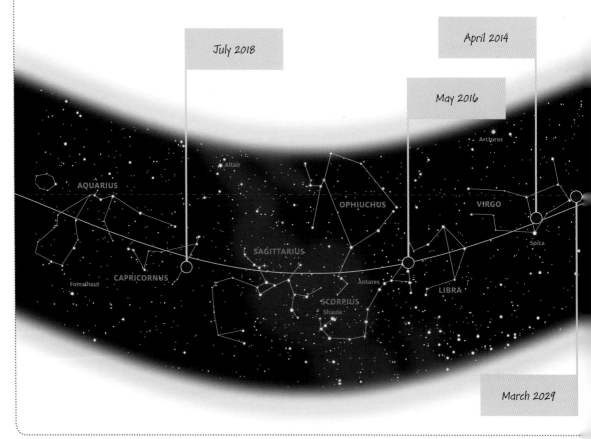

# MARS

## The view from Earth

Mars takes 687 days (just under 2 years) to orbit the Sun. Earth takes one year, so every 25–26 months it overtakes Mars "on the inside". This is known as "opposition"; for about 6 weeks Mars is at its brightest and closest to Earth – and easiest to see. You can't see Mars when it is on the other side of the Sun from Earth ("conjunction"), which also happens every 25–26 months.

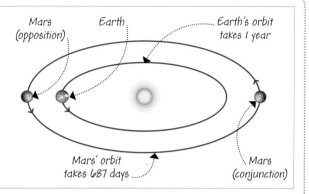

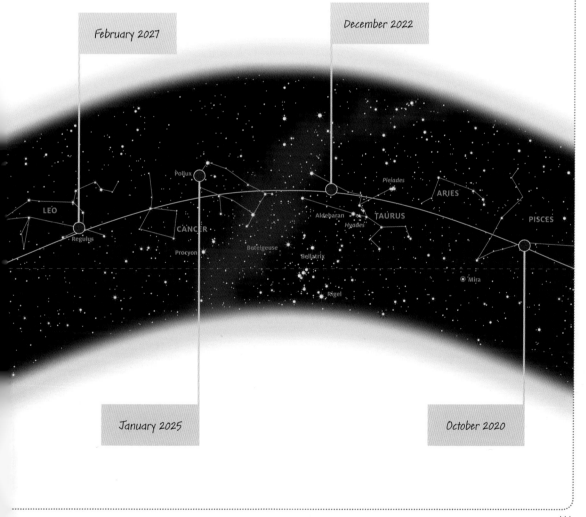

February 2027

December 2022

January 2025

October 2020

# A Quick Guide to **Mars**

Known as the Red Planet due to its rust-red colour, Mars is about half the size of Earth and, like our home planet, has seasons and polar ice caps. Evidence from space probes suggests that water once flowed across the surface, but today it is dry and barren, with huge extinct volcanoes, massive canyons, and vast rock-strewn plains.

## Orbit

Mars takes 687 days to complete its elliptical orbit of the Sun; a year on Mars is nearly twice as long as a year on Earth. The distance from the Sun varies by up to 42.6 million km (26.5 million miles), and this, combined with its fairly heavily tilted spin axis, causes significant seasonal variation on the planet's surface.

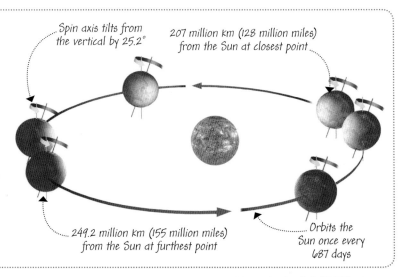

Spin axis tilts from the vertical by 25.2°

207 million km (128 million miles) from the Sun at closest point

249.2 million km (155 million miles) from the Sun at furthest point

Orbits the Sun once every 687 days

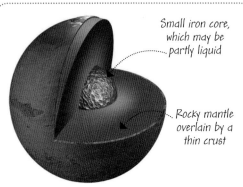

Small iron core, which may be partly liquid

Rocky mantle overlain by a thin crust

## Structure

Mars is similar to Earth in structure, with a distinct core, mantle, and crust. Its atmosphere – mainly carbon dioxide – is extremely thin. Winds periodically cause dust storms that cover the whole planet.

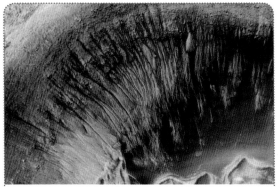

## Martian gullies

This Martian crater created by an ancient asteroid impact is about 7km (4.4 miles) across. The narrow gullies eroded into its walls are thought to have been formed by flowing water in the distant past.

# MARS

## Mars profile

Mars is the fourth planet out from the Sun. With a diameter of 6,792km (4,221 miles) it is the second smallest planet in the Solar System.

**Average distance from the Sun**
227.9 million km (141.6 million miles)

**Orbits the Sun once every**
687 days

**Rotates once every**
24 hours 37 minutes 23 seconds

**Greatest apparent magnitude**
-2.9 (see p.51)

**Furthest from Earth**
401.3 million km (249.4 million miles)

**Nearest to Earth**
54.5 million km (33.9 million miles)

**Greatest angular size in sky**
25.1 arcseconds (see p.12)

**Number of moons**
2

**Size comparison**

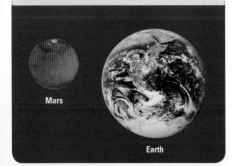

Polar ice cap of frozen water and carbon dioxide

Rust-red surface with an average temperature of -63°C (-81°F)

113

OBSERVATION TECHNIQUES

# Observing the **Orion Nebula**

As you've seen, Orion is one of the most spectacular constellations in the night sky, but nestling within it is a massive glowing cloud of gas and dust called the Orion Nebula – a fantastic sight through binoculars.

## Locating the Nebula

Start by finding Orion's belt with your naked eye (see p.60), and look down from the middle star – or look upwards from the middle star if you are observing from the southern hemisphere. You should see three stars in a line, forming what is known as Orion's sword; the middle one is rather fuzzy. Aim your binoculars at the sword region and slowly sweep them around until you find the middle "star" of the sword. It will not appear as a single point of light like the other stars, but as a much larger hazy patch. It is not in fact a star at all: this is the Orion Nebula.

**Orion from the northern hemisphere**

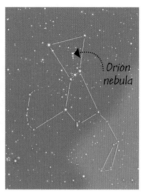

**Orion from the southern hemisphere**

## Through a lens

Viewed through binoculars the nebula has a slight greenish or blue-green hue. With a telescope, more colour may become apparent; the image opposite shows the nebula through a large telescope.

## What's in a nebula?

The Orion Nebula is a massive cloud of gas and dust within our own Milky Way galaxy, where new stars and planets are being formed. About 1,350 light years away from Earth – light from the nebula takes 1,350 years to reach us – and around 24 light years across, it is the closest region of the Milky Way to Earth in which new stars are being formed. The total mass of gas in the nebula is estimated to be greater than that of 2,000 suns. The nebula glows because of the light from many young stars embedded within it.

**Orion Nebula M42 through the Hubble Telescope**

# A Quick Guide to **Light and Dark Nebulae**

Scattered between the stars of the Milky Way are clouds of gas and dust. Some are dense enough to reflect or block starlight, forming visible structures known as nebulae. The different nebulae are among the most beautiful sights in the galaxy.

## Emission nebulae

These brightly coloured clouds are illuminated from within by newborn stars. Intense radiation from the stars causes hydrogen atoms in the cloud to ionize (lose electrons); when the electrons recombine with hydrogen nuclei, red light is emitted. Green or blue light indicates that other elements – oxygen, helium, or nitrogen – are present. Emission nebulae include star-forming regions and glowing clouds of debris ejected by dying stars.

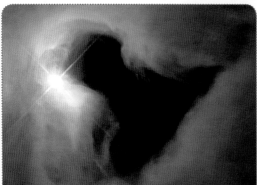

## Reflection nebulae

Young stars are often surrounded by dust and gas left over from the nebula that gave birth to them. These grains of dust may reflect the bluish light of the hot young stars, forming a blue reflection nebula.

## Dark nebulae

Cool clouds of dust and molecular hydrogen can block the light from more distant stars, forming silhouettes known as dark nebulae. These nebulae can collapse under their own gravity and become star-forming regions.

The Bubble Nebula NGC 7635

# KEY CONSTELLATIONS

## Observing **Ophiuchus**

This large constellation is rich in globular star clusters, and a fine sight through binoculars. Although it lies partly in the band of sky called the zodiac, it has never been counted as a "sign of the zodiac".

> **Remember** Within the constellation area of Serpens Cauda adjoining Ophiuchus is the Eagle Nebula, or M16, about the size of the full Moon. If you train binoculars on it you may be able to spot a star cluster in M16, which appears as a hazy ball of light.

### Ophiuchus, the Serpent-bearer

Ophiuchus is closely associated with the two-part constellation of Serpens – Serpens Caput (Head of the Serpent) and Serpens Cauda (Tail of the Serpent). The figure of Ophiuchus is depicted holding the snake. Its brightest star is Rasalhague, a white giant with a smaller companion, and interesting star clusters are located in different parts – near one foot, around one shoulder (both these areas are in the Milky Way), and also around the centre of the constellation.

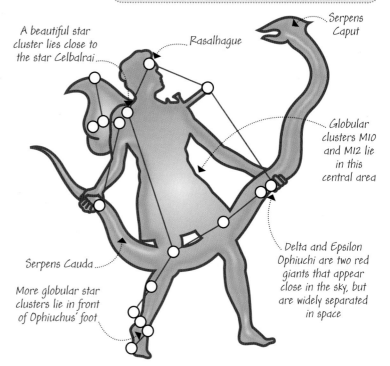

A beautiful star cluster lies close to the star Celbalrai

Rasalhague

Serpens Caput

Globular clusters M10 and M12 lie in this central area

Delta and Epsilon Ophiuchi are two red giants that appear close in the sky, but are widely separated in space

Serpens Cauda

More globular star clusters lie in front of Ophiuchus' foot

## **Northern hemisphere** locators

From mid-latitudes (around 40°N) most of Ophiuchus can be seen above the northern horizon around midnight during May and June. By September it can be seen low in the southwest earlier in the evening.

**1 March, 3am**
Looking southeast

**1 June, midnight**
Looking south

**1 September, 9pm**
Looking southwest

# OPHIUCHUS

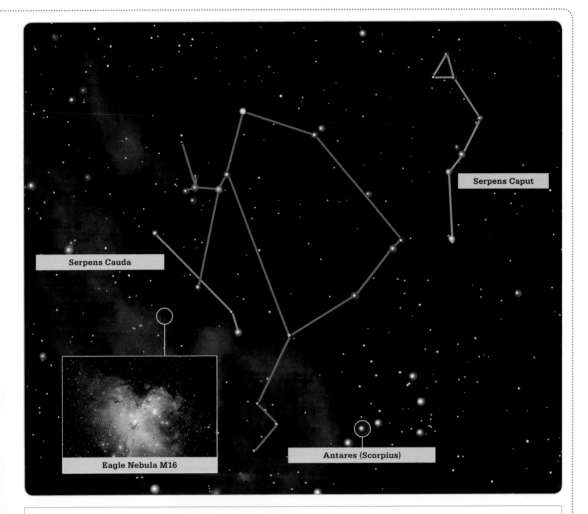

## Southern hemisphere locators

From mid-latitudes Ophiuchus is visible from around March in the east, but only in the early hours of the morning. By June it is high in the north or overhead at midnight, while by September it can be seen in the west in the evening.

**1 March, 3am**
Looking east

**1 June, midnight**
Looking high in north

**1 September, 9pm**
Looking west

KEY CONSTELLATIONS

# Observing **Crux**

You met Crux (the Southern Cross) earlier when you used it to "star hop" to neighbouring constellations and also to find the south celestial pole. The constellation itself lies in a bright area of the Milky Way and contains some interesting objects for binocular or naked-eye viewing.

## Crux, the Southern Cross

Crux is the smallest constellation: from Earth it's just a closed hand's width across. One of its highlights – in addition to the brilliant stars of the cross – is a glittering star cluster, the Jewel Box. Binoculars reveal that it contains several dozen blue-white stars and a contrasting red supergiant. Next to Crux is the Coalsack Nebula, a vast dust cloud.

> **Remember** Crux is one of the most easily recognized constellations, with a familiar shape – the unmistakable arrangement of its stars is displayed on the national flags of New Zealand, Australia, Samoa, Papua New Guinea, and Brazil.

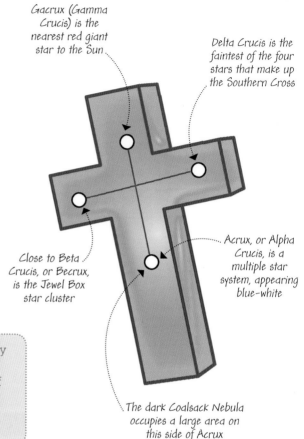

*Gacrux (Gamma Crucis) is the nearest red giant star to the Sun*

*Delta Crucis is the faintest of the four stars that make up the Southern Cross*

*Close to Beta Crucis, or Becrux, is the Jewel Box star cluster*

*Acrux, or Alpha Crucis, is a multiple star system, appearing blue-white*

*The dark Coalsack Nebula occupies a large area on this side of Acrux*

## **The Coalsack** dark nebula

This large smudgy cloud of interstellar dust is silhouetted against the Milky Way. It spans the width of 12 full Moons – so on dark nights it is easy to spot. Here, the star Acrux is to the right of the nebula, and Beta Crucis is above it.

# CRUX

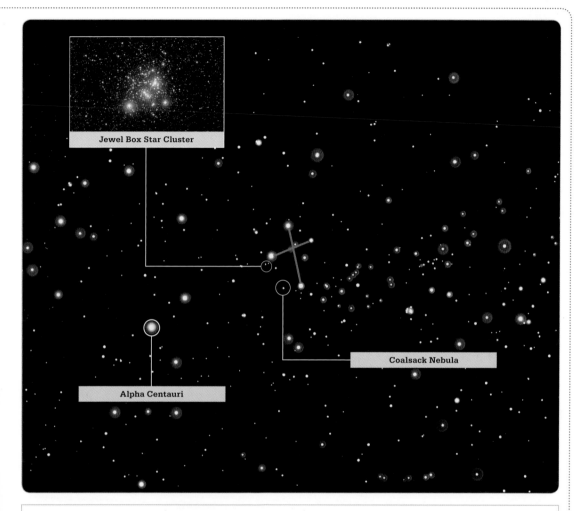

## Southern hemisphere locators

From mid-latitudes the whole of Crux is visible in the southern sky for much of the night throughout the year. Part of the constellation will sometimes dip below the horizon – for example during November evenings.

**1 December, 2am**
Looking south

**1 April, midnight**
Looking high in south

**1 August, 10pm**
Looking south

KEY CONSTELLATIONS

## Observing **Sagittarius**

This prominent zodiacal constellation lying in the direction of the centre of the Milky Way is home to numerous star clusters and bright nebulae, which are a rewarding sight through binoculars. Sagittarius is most easily viewed from the southern hemisphere, but on summer nights it can be well worth taking a look at it from the northern hemisphere too.

> **Remember** An impressive sight to seek out with binoculars in Sagittarius is the Lagoon Nebula (M8), an emission nebula; it lies above the spout of the teapot asterism. Further north lies the Omega – or Swan – Nebula, appearing pale green.

## Sagittarius, the Archer

The constellation figure of Sagittarius classically shows a centaur (half human, half horse) with an archer's bow. To find it in the sky, it may help to first identify the teapot asterism – a pattern of stars at the centre of the constellation that resembles a teapot. The area around the top of the teapot is the best place to look with binoculars to see bright nebulae and star clusters.

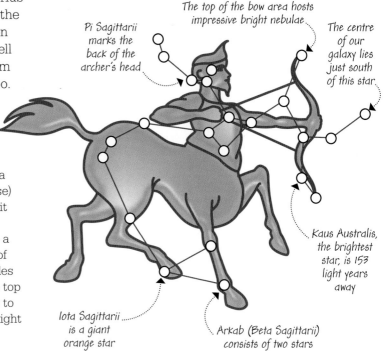

Pi Sagittarii marks the back of the archer's head

The top of the bow area hosts impressive bright nebulae

The centre of our galaxy lies just south of this star

Kaus Australis, the brightest star, is 153 light years away

Iota Sagittarii is a giant orange star

Arkab (Beta Sagittarii) consists of two stars

## **Northern hemisphere** locators

From mid-latitudes (40°N) the constellation of Sagittarius can be viewed between May and September, looking towards the southern horizon. During July, most of the constellation is above the horizon at around midnight.

**1 May, 2am**
Looking south

**1 July, midnight**
Looking south

**1 September, 10pm**
Looking south

# SAGITTARIUS

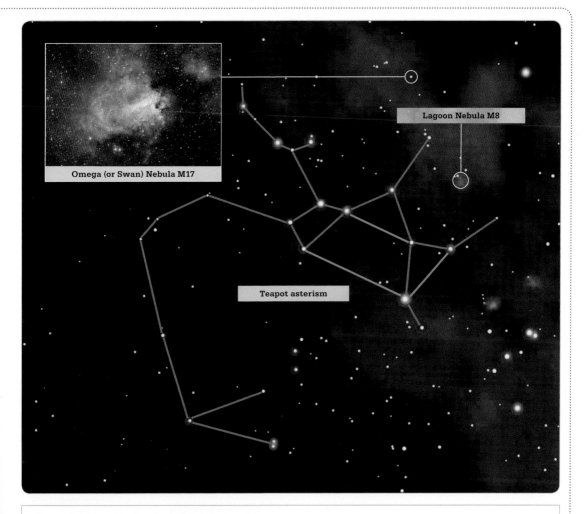

## Southern hemisphere locators

From mid-latitudes (40°S), Sagittarius is visible from March in the early hours of the morning, looking east. By July it is high in the north or overhead around midnight. In October it can still be seen in the evening looking west.

**1 April, 2am**
Looking east

**1 July, midnight**
Looking high in north

**1 October, 10pm**
Looking west

# KEY CONSTELLATIONS

## Observing **Carina**

A southern constellation that lies partly within the band of the Milky Way, Carina contains a bright nebula, one exceptionally bright star, and several star clusters. Unfortunately, Carina cannot be seen from the northern hemisphere unless you are close to the equator.

### Carina, the Keel

The constellation of Carina was once part of the larger figure of Argo Navis, the ship of the Argonauts. In the 18th century the keel, sails, and stern were split into Carina, Vela (see p.158), and Puppis. Carina retained many of the original ship's most spectacular features, such as the intense white star Canopus and the bright Eta Carinae Nebula.

> **Remember** Most of Carina's star clusters, which are recommended for binocular viewing, lie near the Carina Nebula, which is easy to spot with the naked eye. An exception is the Garden Sprinkler Cluster, which lies further over, towards Canopus.

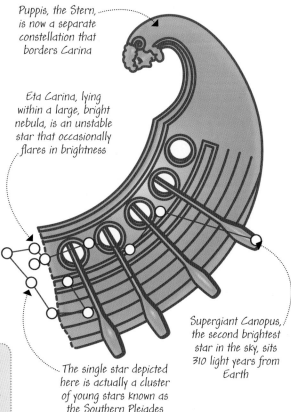

*Puppis, the Stern, is now a separate constellation that borders Carina*

*Eta Carina, lying within a large, bright nebula, is an unstable star that occasionally flares in brightness*

*Supergiant Canopus, the second brightest star in the sky, sits 310 light years from Earth*

*The single star depicted here is actually a cluster of young stars known as the Southern Pleiades*

### Eta Carinae Nebula NGC 3372

Carina's centrepiece is a large bright cloud of gas and dust 7,500 light years away. The Eta Carinae Nebula, or Great Nebula, is an important star-forming region in the Milky Way, and an excellent target for binocular viewing.

# CARINA

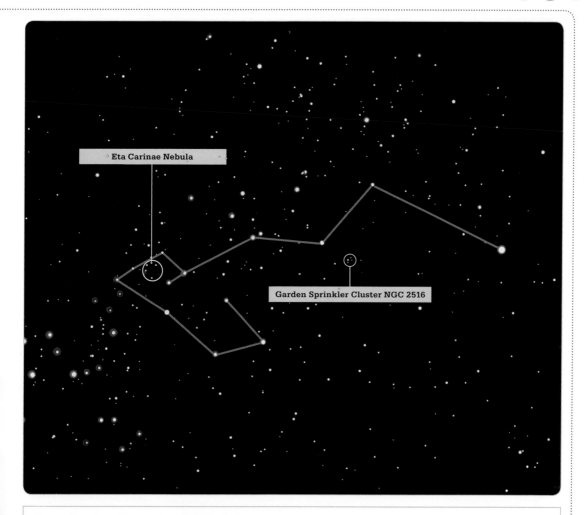

## Southern hemisphere locators

From mid-latitudes the constellation of Carina is at least partly visible in the southern sky on every night of the year, although a portion of it does sometimes dip below the horizon – for example around midnight in August.

**1 October, 2am**
Looking south

**1 February, midnight**
Looking high in south

**1 June, 10pm**
Looking south

# KEY CONSTELLATIONS

## Observing **Andromeda**

This northern constellation may not be one that leaps out of the sky, but it does contain some good objects for binocular viewing. Of note is the distant Andromeda galaxy (see p.128) – but first get to know Andromeda itself, and look at some of the closer objects in it.

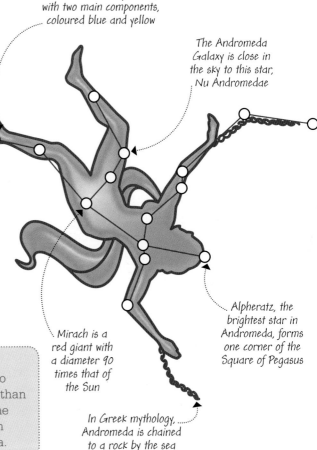

*Almach is a multiple star, with two main components, coloured blue and yellow*

*The Andromeda Galaxy is close in the sky to this star, Nu Andromedae*

*Alpheratz, the brightest star in Andromeda, forms one corner of the Square of Pegasus*

*Mirach is a red giant with a diameter 90 times that of the Sun*

*In Greek mythology, Andromeda is chained to a rock by the sea*

## Andromeda, the Princess

The figure depicts the daughter of the mythical Queen Cassiopeia, who has her own constellation (see p.104). It consists of several strings of relatively faint stars joined to an uneven line of brighter stars running from Andromeda's left foot to her head. Alpheratz (at her head) marks one corner of the Square of Pegasus, a large, easy-to-spot nearby asterism.

> **Remember** The prominent Square of Pegasus surrounds a dark area of sky, so initially you may find it easier to locate than Andromeda itself. Once you've found the Great Square use the star it shares with the constellation to identify Andromeda.

## **Northern hemisphere** locators

At mid-latitudes the constellation is visible at some point during the night for most of the year. The best seasons for viewing it are autumn and early winter, when Andromeda is high in the south or overhead in late evening.

**1 June, 3am**
Looking east

**1 October, midnight**
Looking high in south

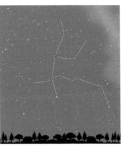

**1 February 9pm**
Looking west

# ANDROMEDA

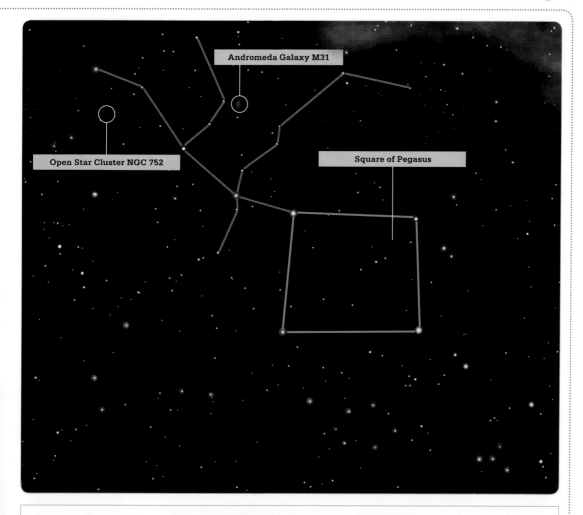

## Open Star Cluster NGC 752

This star cluster is found in the area below Andromeda's left leg. It consists of around 70 stars in a loose group, and looks great through binoculars, appearing next to a curving chain of brighter, coloured stars to its south.

OBSERVATION TECHNIQUES

# Observing the **Andromeda Galaxy**

While studying Andromeda and the Square of Pegasus you might like to use your binoculars to seek out a special object – the Andromeda Galaxy – the nearest large galaxy outside our own Milky Way.

## Locating the galaxy

Evenings from October to December, in the northern hemisphere or low latitudes in the southern, are the best times for locating the galaxy. Find Andromeda and the Square of Pegasus just using your eyes. Next, through binoculars, locate the bright star Alpheratz at one corner of the square.

Sweep your binoculars along the figure's body. When you get to the second star, Mirach, turn right along Andromeda's leg towards the "thigh" area. At the second star in the thigh, turn "right" a little way. You should see a large fuzzy elliptical patch of light – this is the Andromeda galaxy.

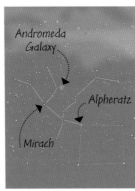

**Andromeda from the northern hemisphere**

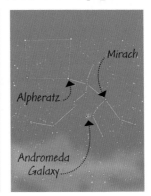

**Andromeda from the southern hemisphere**

### Through binoculars

The stretched-out misty ellipse is larger than the Moon – and this is just the galaxy's central area. The whole galaxy is about twice as big, but its outer parts are too faint to see through binoculars.

## Looking through time

On a very dark night with no light pollution the Andromeda Galaxy is the most distant object that can be seen with the naked eye – it's over 2.5 million light years away, so you see the galaxy as it appeared 2.5 million years ago. A spiral galaxy, like our own Milky Way Galaxy, it contains about a trillion stars and is somewhat larger than the Milky Way, but of comparable mass. Unlike most other galaxies it is moving towards our own, and in about 3.75 billion years' time, the two are expected to merge.

# ANDROMEDA GALAXY

**Andromeda Galaxy M31**

# A Quick Guide to **Galaxies**

Galaxies are vast, spinning collections of stars, gas, dust, and dark matter, held together by the force of gravity. Astronomers believe there are at least 100 billion galaxies in the Universe. They vary a great deal in size and shape, from dwarfs with just a few million stars to giants with over 100 trillion.

## Types of galaxies

Galaxies are classified into just a few main types, depending on the shape we see from Earth. Spiral galaxies have a central hub of stars surrounded by curving spiral arms. In barred spiral galaxies, a straight bar runs across the centre, joining the spiral arms. Elliptical galaxies are simple ball shapes, and irregular galaxies have no clear shape.

**Elliptical galaxy**

**Spiral galaxy**

**Barred spiral galaxy**

**Irregular galaxy**

# GALAXIES

**Spiral galaxy NGC 4414**

## Active galaxies

Astronomers believe that most galaxies revolve around a central supermassive black hole. Most black holes appear to be dormant, but in certain galaxies – "active" galaxies – matter is falling into the black hole. As it spirals inward, the matter is torn apart by gravity and whirls around the galactic heart to form a ring of superheated dust and gas. Two vast jets of particles blast out from this ring, travelling close to the speed of light and stretching thousands of light years into space. Active galaxies are classified into various types according to their appearance and the mix of radiation they emit.

# KEY CONSTELLATIONS

## Observing Leo

The zodiacal constellation of Leo is one of the easiest to recognize, because its outline really does bear a resemblance to a lion. Highlights for binocular viewing include a few more galaxies as well as some of Leo's brighter stars.

**Remember** The bright star Regulus in Leo can be found by "starhopping" from the Plough (see p.46). Once you've found it, you should also be able to see the Sickle asterism, which looks like a back-to-front question mark.

## Leo, the Lion

The constellation figure of Leo depicts a crouching lion. At the front of the figure, an asterism called the Sickle traces the lion's chest and head. Good objects for binocular viewing include several stars of the Sickle and a region around the lion's back right leg that contains three faint spiral galaxies. The Leonid meteors radiate from the region of the Sickle every November (see p.74).

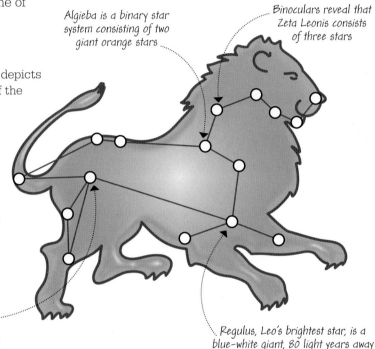

Algieba is a binary star system consisting of two giant orange stars

Binoculars reveal that Zeta Leonis consists of three stars

Just below and behind this star lie three spiral galaxies – M65, M66, and NGC3628

Regulus, Leo's brightest star, is a blue-white giant, 80 light years away

## Northern hemisphere locators

From mid-latitudes (around 40°N) the constellation of Leo becomes visible from early November in the eastern sky – but only in the early morning hours. By March it is prominent high in the south at around midnight.

**1 December, 2am**
Looking east

**1 March, midnight**
Looking high in south

**1 June, 10pm**
Looking west

# LEO

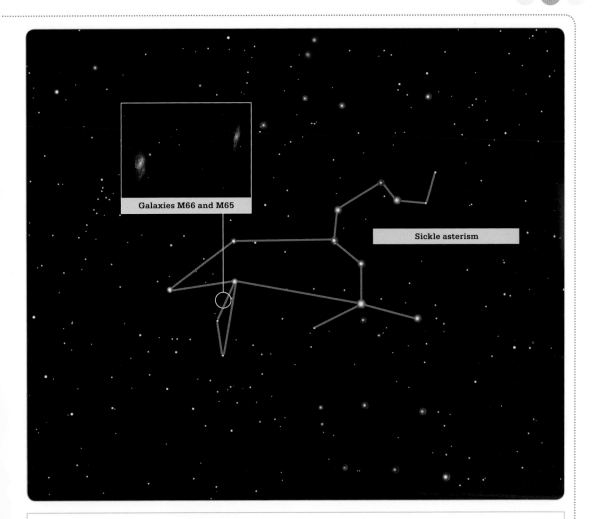

## Southern hemisphere locators

From mid-latitudes (around 40°S), the constellation can be seen — but upside down — low in the north in February and March at around midnight. By May Leo is low in the northwestern sky earlier in the evening.

**1 January, 2am**
Looking northeast

**1 March, midnight**
Looking north

**1 May, 10pm**
Looking northwest

# KEY CONSTELLATIONS

## Observing **Dorado**

From the southern hemisphere, it's easy to see a galaxy outside our own: find the constellation of Dorado and look for the Large Magellanic Cloud (LMC), which is an irregular satellite galaxy of the Milky Way. It is visible to the naked eye, but the LMC – as well as a number of other objects in Dorado – is also worth examining through binoculars.

**Remember** To locate Dorado and the Large Magellanic Cloud, first establish the direction of the south celestial pole (see p.44). The "mouth" of the fish and the LMC lie roughly between the pole and Canopus, the brightest star in this part of the sky.

## Dorado, the Dolphinfish

The constellation figure depicts a fish – of a type found only in tropical waters – swimming southwards. Several of its six main stars are worth looking at through binoculars, but the most interesting object is the nearby Large Magellanic Cloud. Classed as a dwarf galaxy, this is named after the 16th century explorer Ferdinand Magellan. Although some 170,000 light years away, it spans 20 full Moon diameters in the sky, and at first sight looks like a detached part of the Milky Way. Binoculars bring into view numerous nebulae and star clusters.

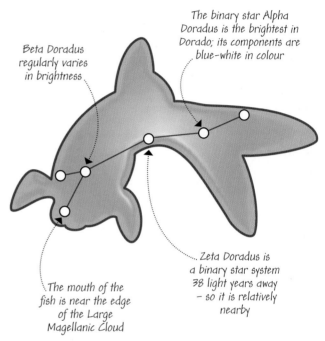

The binary star Alpha Doradus is the brightest in Dorado; its components are blue-white in colour

Beta Doradus regularly varies in brightness

Zeta Doradus is a binary star system 38 light years away – so it is relatively nearby

The mouth of the fish is near the edge of the Large Magellanic Cloud

## **Tarantula Nebula** NGC 2070

This star-forming nebula is near one edge of the Large Magellanic Cloud. At its heart, a cluster of new-born stars can be seen with binoculars. Loops of gas in the nebula resembling a spider's legs inspired its popular name.

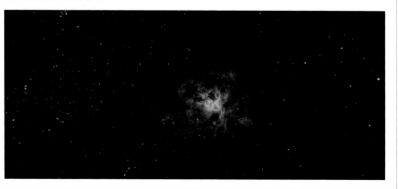

# DORADO

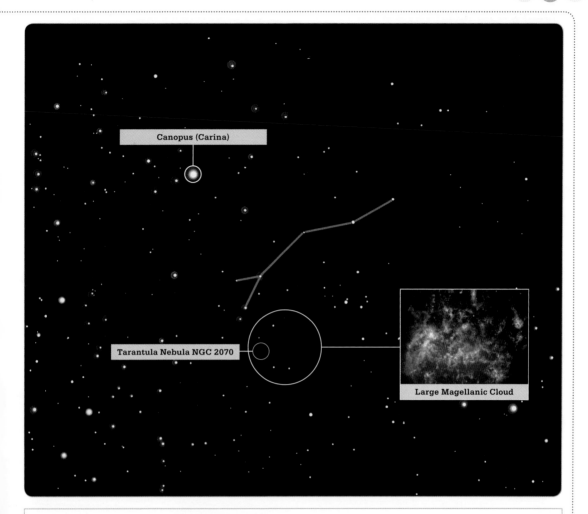

## Southern hemisphere locators

From mid-latitudes (around 40°S) the constellation of Dorado is at least partly visible in the southern sky every night of the year. Parts of it sometimes dip below the horizon – for example during June and July evenings.

**1 August, 2am**
Looking southeast

**1 December, midnight**
Looking high in south

**1 April, 10pm**
Looking southwest

# Observing **Mercury**

Mercury, the closest planet to the Sun, is a tricky object to observe because it never strays far from the Sun in the sky. But from time to time it can be seen as a pinprick of light near the horizon at sunset or dawn. Use of binoculars may reveal its phases.

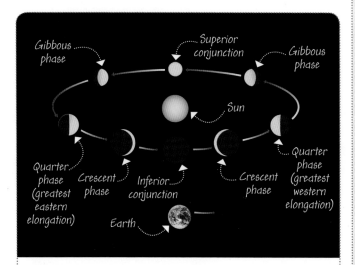

## The view from Earth

Mercury orbits the Sun much faster than Earth and roughly every 116 days overtakes Earth "on the inside". At the point of each overtake Mercury moves between Earth and the Sun (inferior conjunction) and cannot be seen. But within a few weeks, and for a few weeks thereafter, it is visible low in the east before dawn (morning apparitions). Subsequently for a few weeks it moves out of view to the other side of the Sun (superior conjunction). The planet then reappears for a few more weeks, this time low in the west just after sunset (evening apparitions). Finally, we lose sight of Mercury as it completes its orbit and overtakes Earth again.

**Tip** The chances of an observation depend on your latitude, the time of year, and other factors related to Mercury's orbit. But in a clear sky there's a reasonable chance you will see Mercury as a small bright spot very close to the horizon. Try not to confuse it with Venus, which is much brighter and more dazzling.

**LOOK EAST before dawn (morning apparitions)**

| Date | Greatest elongation |
| --- | --- |
| Mar 2014 | 14 Mar |
| Jul 2014 | 12 Jul |
| Oct–Nov 2014 | 1 Nov |
| Feb 2015 | 24 Feb |
| Jun 2015 | 24 Jun |
| Oct 2015 | 16 Oct |
| Feb 2016 | 6 Feb |
| May–Jun 2016 | 5 Jun |
| Sep–Oct 2016 | 28 Sep |
| Jan 2017 | 19 Jan |
| May 2017 | 17 May |
| Sep 2017 | 12 Sep |
| Dec 2017–Jan 2018 | 1 Jan |
| Apr–May 2018 | 29 Apr |
| Aug–Sep 2018 | 26 Aug |
| Dec 2018 | 15 Dec |
| Apr 2019 | 11 Apr |
| Aug 2019 | 9 Aug |
| Nov–Dec 2019 | 28 Nov |
| Mar 2020 | 24 Mar |
| Jul 2020 | 22 Jul |
| Nov 2020 | 10 Nov |
| Mar 2021 | 6 Mar |
| Jun–Jul 2021 | 4 Jul |
| Oct 2021 | 25 Oct |
| Feb 2022 | 16 Feb |
| Jun 2022 | 16 Jun |
| Oct 2022 | 8 Oct |
| Jan–Feb 2023 | 30 Jan |

# MERCURY

| LOOK WEST after sunset (evening apparitions) | |
|---|---|
| **Date** | **Greatest elongation** |
| Jan–Feb 2014 | 31 Jan |
| May 2014 | 25 May |
| Sep 2014 | 21 Sep |
| Jan 2015 | 14 Jan |
| May 2015 | 7 May |
| Aug–Sep 2015 | 4 Sep |
| Dec 2015–Jan 2016 | 29 Dec |
| Apr 2016 | 18 Apr |
| Aug 2016 | 16 Aug |
| Dec 2016 | 10 Dec |
| Mar–Apr 2017 | 1 Apr |
| Jul–Aug 2017 | 29 Jul |
| Nov 2017 | 23 Nov |
| Mar 2018 | 15 Mar |
| Jul 2018 | 11 Jul |
| Oct–Nov 2018 | 6 Nov |
| Feb–Mar 2019 | 27 Feb |
| Jun 2019 | 23 Jun |
| Oct 2019 | 20 Oct |
| Feb 2020 | 10 Feb |
| May–Jun 2020 | 4 Jun |
| Sep–Oct 2020 | 1 Oct |
| Jan 2021 | 24 Jan |
| May 2021 | 17 May |
| Sep 2021 | 14 Sep |
| Jan 2022 | 7 Jan |
| Apr–May 2022 | 29 Apr |
| Aug–Sep 2022 | 27 Aug |
| Dec 2022 | 21 Dec |
| Apr 2023 | 11 Apr |

**Mercury (left) with Venus at sunset over Chile, October 2011**

## Locating Mercury

Mercury is easier to see from tropical and subtropical latitudes, but to view it choose a date between greatest elongation, when it is at its maximum distance from the Sun and most easily spotted, and about a week later for morning apparitions, or a week earlier for evening apparitions. Once you've located Mercury, try examining it through binoculars. You may see that its disc is not fully illuminated but rather half-illuminated or even in a slightly gibbous phase, because Mercury has phases like that of the Moon.

# A Quick Guide to **Mercury**

Mercury is one of the more extreme worlds in the Solar System. Due to a unique relationship between its orbit and rate of rotation, most parts of its surface are alternately scorched by the Sun, then plunged into darkness and extreme cold for months at a time. The surface is covered in impact craters and some ancient lava flows.

## Orbit

Mercury's 88-day orbit is highly elliptical: at its greatest distance from the Sun it is more than 50 per cent further away than at its closest point. Mercury spins on its axis once every 58.7 Earth days, so it makes three spins for every two orbits. As a result, on Mercury 176 Earth days pass between one sunrise and the next.

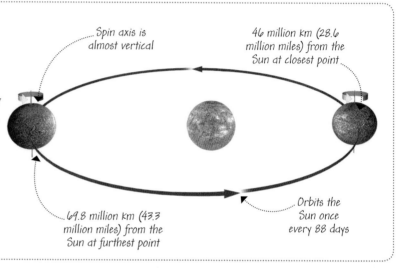

Spin axis is almost vertical

46 million km (28.6 million miles) from the Sun at closest point

69.8 million km (43.3 million miles) from the Sun at furthest point

Orbits the Sun once every 88 days

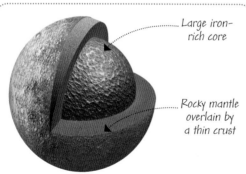

Large iron-rich core

Rocky mantle overlain by a thin crust

## Structure

Mercury's density is known to be high, so its core must be relatively large and iron-rich. Around the core are a rocky mantle and crust. It is too small and (periodically) hot to retain a significant atmosphere.

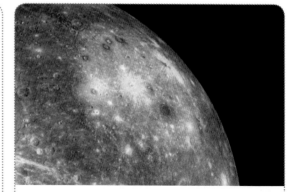

## Caloris Basin

The circular area with a yellowish hue in this colour-enhanced image is known as the Caloris Basin. This wide depression is about 1,500km (930 miles) across, and was created by an ancient asteroid impact.

# MERCURY

## Mercury profile

Mercury is the closest planet to the Sun. With a diameter of 4,879km (3,032 miles) it is also the smallest planet in the Solar System.

**Average distance from the Sun**
57.9 million km (36 million miles)

**Orbits the Sun once every**
88 days

**Rotates once every**
58 days 15 hours 36 minutes

**Greatest apparent magnitude**
-1.9 (see p.51)

**Furthest from Earth**
221.9 million km (138 million miles)

**Nearest to Earth**
77.3 million km (48 million miles)

**Greatest angular size in sky**
13 arcseconds (see p.12)

**Number of moons**
0

**Size comparison**

Surface temperature varies from up to 430°C (806°F) during the day down to -180°C (-292°F) at night

Heavily scarred and cratered surface

OBSERVATION TECHNIQUES

# Observing a Comet

The appearance, for a few weeks or months, of a bright comet in the night sky is a fairly rare astronomical treat. When one of these exotic visitors does appear, spending an hour or two observing it, whether with binoculars or just the naked eye, can be a rewarding experience.

## When to observe a comet

Comets are chunks of ice, dust, and rock that orbit the Sun. As they draw close to the Sun, some develop bright clouds of material around themselves and streaming tails, which makes for a fine sight in the sky. The bad news is that spectacular or "Great" comets are infrequent and unpredictable: there are perhaps one or two per decade. A few return regularly and reliably put on a good show, but the most famous – Halley's Comet – will not be seen again until 2061. Faint comets are much more common, but a "Great" comet will be easier to spot. If one is reported in the media, make sure you see it, even if it means rising at 3am.

# COMETS

## Observation technique

A comet does not streak across the sky like a shooting star (see p.74). Look instead for a fuzzy area of light – the coma – which is stationary against the starry background; over several nights you will see that it moves slowly across the starscape. One or more wispy tails may extend from the coma. It is worth observing a bright comet both with the naked eye – to admire the full sweep of the tail(s) – and also through binoculars, which allow you to see extra subtle details in the tails, such as kinks, clumps, and streamers.

**Comet Hale-Bopp, July 1997**

# A Quick Guide to **Comets**

Comets are spinning chunks of ice, dust, and rock. They are anything from about 100m (328ft) to 40km (25 miles) across, and generally occupy remote parts of the Solar System. When a comet enters the inner Solar System, it can become a noticeably bright object in the sky.

## Cometary orbits and tails

Comets seen in the inner Solar System are either regular visitors – returning at intervals of anything from a few years to thousands of years – or "once only" tourists. All follow stretched elliptical orbits around the Sun. As a comet nears the Sun, the Sun's heat causes it to release gas, dust, and ice particles into space. These create a bright fuzzy cloud round the comet's nucleus (central solid part) called the coma. Radiation and streams of particles emitted by the Sun also push on the cloud of ejected material, typically forming two comet "tails". One consists of dust and the other, usually fainter and straighter, consists of ions (charged gas particles).

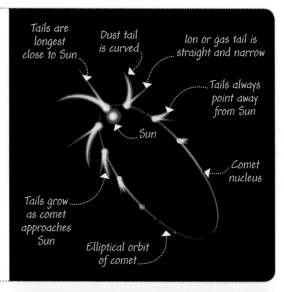

## Comet Pan-STARRS

This comet put on a little show for skywatchers in the early months of 2013, although it never became very bright. It is not expected to reappear for thousands of years – if ever.

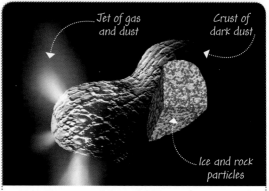

## Nucleus of a comet

A comet's nucleus has an irregular shape and consists of a mixture of ice, dust, and rock particles, with a thin surface layer of dust. As it is warmed by the Sun, jets of gas and dust start to spew from its surface.

Comet McNaught, also called the Great Comet of 2007

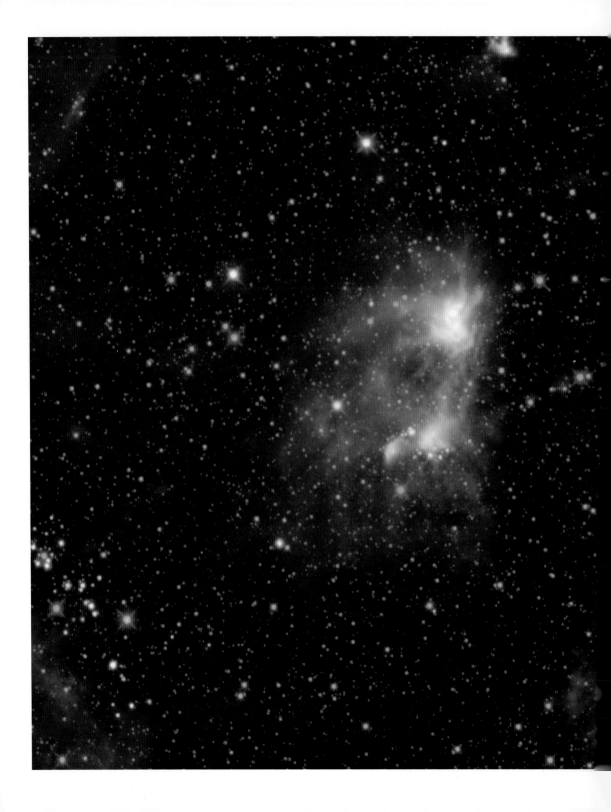

# 3

# Take It Further

The main focus of this final chapter is on some fascinating but rather faint objects in the night sky that you need a telescope to observe clearly. They include Saturn's rings; Jupiter's four largest moons; some planetary nebulae (remains of red giant stars); more galaxies outside our own; and supernova remnants such as the Crab nebula. Once again, you'll meet a handful of new constellations – this time they contain good objects for telescopic viewing. Also included are guides to choosing and using a small telescope, and the best and safest ways to use equipment if you want to observe the Sun. All these areas are best tackled once you have built up some experience in naked eye and binocular astronomy.

**Star-forming nebula (yellow, centre) and Tycho's supernova remnant (red bubble at top right)**

TOOLS AND EQUIPMENT

# Choosing and using **a small telescope**

As your interest in astronomy develops, at some point you may want to start using a telescope. This will open up the night skies to new levels of observation, especially for viewing fainter or more indistinct objects such as the moons of Jupiter, Saturn's rings, and some types of nebulae.

**Tip** Telescopes can have swoppable eyepieces for different magnifications. Bear in mind magnifiers over 300x are not of great use because atmospheric turbulence may outweigh the extra resolving power.

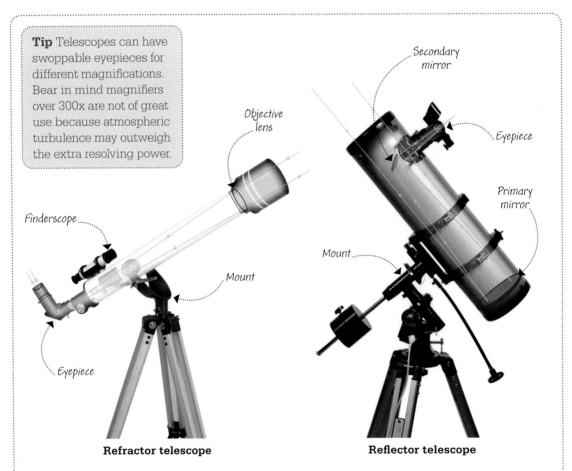

**Refractor telescope**

**Reflector telescope**

## Types of telescope

There are two main designs: refractors collect and focus light using a lens, while reflectors use mirrors. Reflectors tend to be cheaper than refractors and more comfortable to use but also more fragile, and they need more maintenance. A third, more complex design – a catadioptric – uses both lenses and mirrors; it's generally more expensive than an equivalent reflector.

Two important factors to consider when choosing your telescope are its light-gathering ability – which depends on aperture – and its magnifying power. Aperture (the diameter of the main lens or mirror) will affect your ability to view faint objects, so the bigger the better. Magnifying power is governed by several components including the eyepiece.

## Mounts

The mount not only supports the telescope but also helps to track sky objects. There are two main types. An altazimuth pivots in two directions: altitude (up and down) and azimuth (parallel to the horizon); it requires continuous adjustment because sky objects constantly change both altitude and azimuth. Equatorial mounts take account of how objects cross the sky, and once set up (by aligning to one of the celestial poles), objects can be tracked by turning one of the direction controls.

**Tip** An attractive but more expensive option is a motorised, computerised "Go-to" telescope. Once aligned to the night sky, the telescope can automatically locate and move to a multitude of different celestial objects, without any further assistance from the observer.

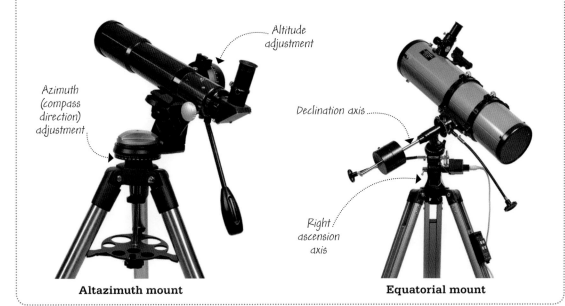

**Altazimuth mount**  **Equatorial mount**

## Finders

All telescopes, even "Go-to" ones, are hard to aim: the typical field of view is 1° so locating objects can be difficult. A finder on the side of the telescope will help. There are two types. Finderscopes have a field of view of 5–8° and use cross-hairs to centre the target. Red dot finders project a small dot onto a transparent piece of plastic; the wider sky is visible, so it is intuitive to use. For both types you need to check the alignment from time to time, by looking at a distant fixed object through the main telescope and then through the finder.

# Observing **Saturn**

Saturn is fairly easy to spot in the night sky once you know where to look. It is always visible to the naked eye when present, but a telescope is needed to make out its famous ring system.

## Tracking Saturn

The easiest way to locate Saturn is to look for it around the time of opposition. The chart below shows all months up to 2030 when Saturn is at, or close to, opposition. For each month, the constellation that Saturn is "in" at the time is also given, so in May 2015 you'll find Saturn in Scorpius.

To spot Saturn around opposition, look into the southern sky (from the northern hemisphere) or overhead at about midnight in the month in question; if you are in the southern hemisphere, look north. Saturn is one of the brighter objects. Astronomical software or a smartphone "app" may also help you locate Saturn.

**Tip** Saturn only moves constellations once every 2–3 years. When it's not in opposition, look for the nearest constellation figure. If you work out when and where to see that, you may be able to spot Saturn too.

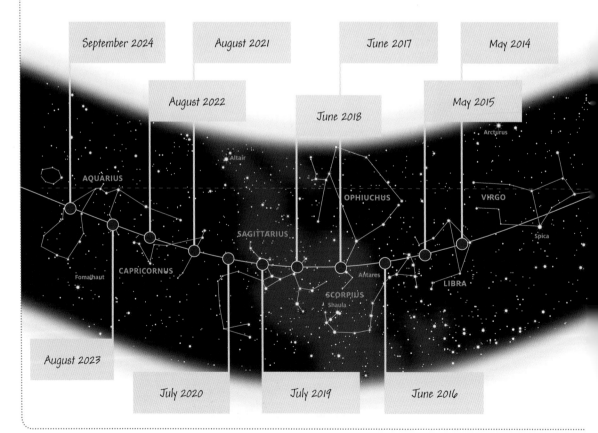

# SATURN

## The view from Earth

Saturn takes nearly 30 years to orbit the Sun once. Earth takes just one year, so it overtakes Saturn "on the inside" roughly every 378 days (just over a year). At this time, "opposition", Saturn is at its closest to Earth and at its brightest. Saturn is not visible when it is on the other side of the Sun from Earth ("conjunction"), which also occurs approximately every 378 days.

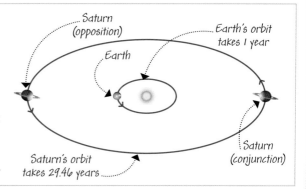

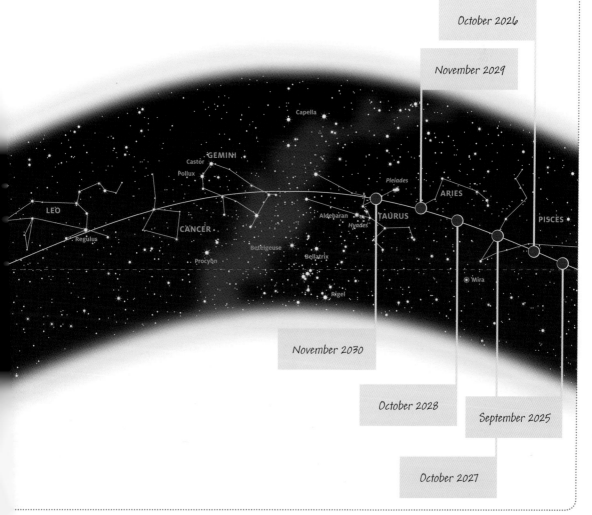

# A Quick Guide to **Saturn**

Saturn is a colossal world consisting almost entirely of the lightest chemical elements, hydrogen and helium. Its magnificent rings are composed of lumps of ice that vary in size from dust-sized specks to icy boulders the size of a car; vast in breadth, on average they are just 20m (66ft) thick. Saturn also has a large family of moons.

## Orbit

Saturn has a markedly elliptical orbit; its distance from the Sun varies by up to 161.9 million km (100.7 million miles). Saturn tilts quite heavily as it spins, causing slow seasonal changes as it moves round the Sun. The axial tilt also means that the view of Saturn's rings, as seen from Earth, continuously changes.

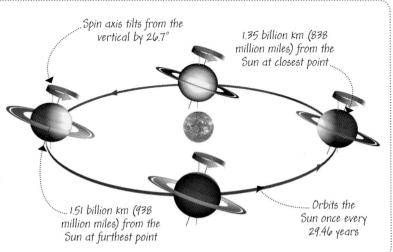

Spin axis tilts from the vertical by 26.7°

1.35 billion km (838 million miles) from the Sun at closest point

1.51 billion km (938 million miles) from the Sun at furthest point

Orbits the Sun once every 29.46 years

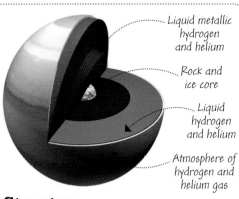

Liquid metallic hydrogen and helium

Rock and ice core

Liquid hydrogen and helium

Atmosphere of hydrogen and helium gas

## Structure

Saturn's centre is thought to be a dense core of rock and ice. Surrounding this are liquid layers, mainly of hydrogen and helium, which gradually merge into an atmosphere of hydrogen and helium gas.

## Saturn's moon Titan

The largest of Saturn's many moons, Titan is the only moon in the Solar System with a dense atmosphere; as on Earth, this consists mostly of nitrogen. On Titan's surface are lakes of liquid methane and ethane.

# SATURN

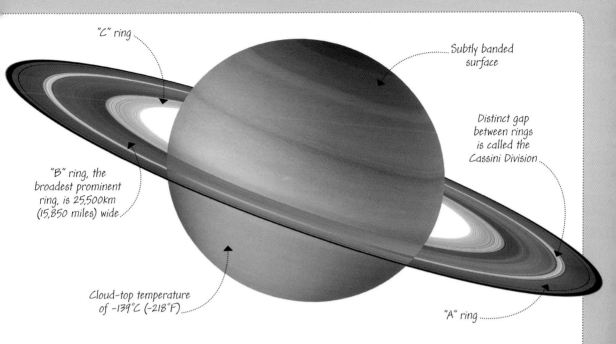

- "C" ring
- Subtly banded surface
- Distinct gap between rings is called the Cassini Division
- "B" ring, the broadest prominent ring, is 25,500km (15,850 miles) wide
- Cloud-top temperature of -139°C (-218°F)
- "A" ring

## Saturn profile

Saturn is the sixth planet out from the Sun. With a diameter of 120,536km (74,898 miles), it is the second largest planet in the Solar System.

| | |
|---|---|
| **Average distance from the Sun** <br> 1.43 billion km (890.8 million miles) | **Greatest angular size in sky** <br> 20.1 arcseconds (see p.12) |
| **Orbits the Sun once every** <br> 29.46 years | **Number of moons** <br> 62 confirmed as of July 2013 |
| **Rotates once every** <br> 10 hours 32 minutes | **Size comparison** |
| **Greatest apparent magnitude** <br> -0.49 (see p.51) | |
| **Furthest from Earth** <br> 1,658.5 million km (1,030.5 million miles) |  <br> Earth — Saturn |
| **Nearest to Earth** <br> 1,195.5 million km (742.8 million miles) | |

151

OBSERVATION TECHNIQUES

# Observing **Jupiter's Galilean Moons**

Now that you're using a telescope, try taking a closer look at Jupiter – see p.68 for a reminder of how to locate it in the night sky. You may be able to pick out some banding and possibly the Great Red Spot (see p.71) on the surface, as well as some of its moons. The four largest were spotted in 1610 by the Italian astronomer Galileo Galilei, who was observing Jupiter through an early telescope – use a small modern telescope to repeat the exercise.

## The Galilean moons

Jupiter has many moons – 67 had been discovered by July 2013 – but the four Galilean moons are much bigger than the rest. Galileo's discovery of Io, Europa, Ganymede, and Callisto was of huge scientific significance, as it assisted in debunking the standard belief of the time that everything in space revolved around Earth.

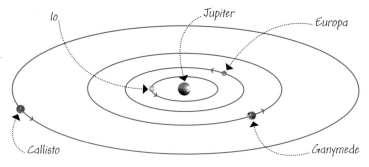

## The changing configuration of Jupiter's moons

If you observe the moons over several nights, you'll see that their configuration changes every night. Galileo also noted this and worked out that it must be because the moons orbit Jupiter, doing so at different orbital periods. Io, the closest to Jupiter, orbits in about 1.8 days; Europa in 3.6 days; Ganymede in 7.2 days; while Callisto takes over 16 days.

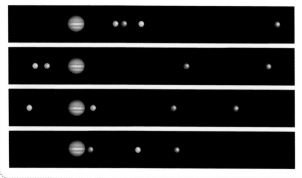

**Moon movements**
The diagram shows how their varied orbits appear to affect the order of the Galilean moons on successive nights. The moons in the top section are (from left to right) Io, Ganymede, Europa, and Callisto.

# JUPITER'S GALILEAN MOONS

1 2 **3**

## Locating the Galilean moons

Aim your telescope at Jupiter and look around it for up to four small white star-like dots, of roughly the same size and brightness in a fairly straight line with Jupiter itself. These are the Galilean moons. They may all be on one side of Jupiter but usually you'll see some to either side, and sometimes one or two will be hidden by the planet.

**Jupiter and the four Galilean moons (left to right): Callisto, Europa, Io, and Ganymede**

KEY CONSTELLATIONS

## Observing **Aquarius**

This large zodiacal constellation can be seen from both hemispheres. It contains a handful of objects that are good targets for a telescope, including two planetary nebulae – glowing shells of gas ejected by red giant stars as they begin to die.

> **Remember** In a clear dark sky, the Helix Nebula is just visible with a small telescope as a pale fuzzy disc, half the width of the full Moon. Around 700 light years away, it is thought to be the closest planetary nebula to Earth.

## Aquarius, the Water Carrier

The figure depicts a young man pouring water from a jar. One planetary nebula, the Saturn Nebula, is located just below a star near the left hand. The other, the Helix Nebula, lies south of the figure. Another notable telescopic object is globular star cluster M2, situated in the vicinity of the head.

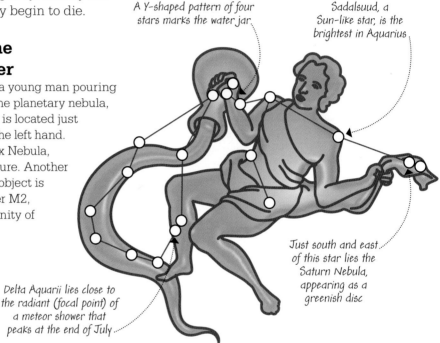

A Y-shaped pattern of four stars marks the water jar.

Sadalsuud, a Sun-like star, is the brightest in Aquarius.

Just south and east of this star lies the Saturn Nebula, appearing as a greenish disc.

Delta Aquarii lies close to the radiant (focal point) of a meteor shower that peaks at the end of July.

## **Northern hemisphere** locators

From mid-latitudes (around 40°N) the constellation of Aquarius is visible above the southern horizon around midnight in August and September. By December it is low in the southwest earlier in the evening.

**1 June, 3am**
Looking southeast

**1 September, midnight**
Looking south

**1 December, 9pm**
Looking southwest

# AQUARIUS

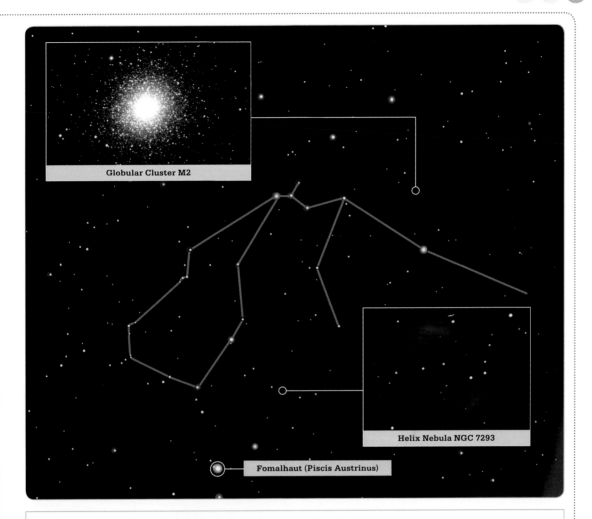

Globular Cluster M2

Helix Nebula NGC 7293

Fomalhaut (Piscis Austrinus)

## Southern hemisphere locators

From mid-latitudes (around 40°S) Aquarius can be seen from June in the northeast in the early morning. By September it is visible high in the north at midnight, and during December in the northwest in the evening.

**1 June, 3am**
Looking northeast

**1 September, midnight**
Looking north

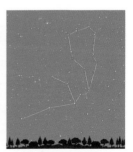

**1 December, 9pm**
Looking northwest

# A Quick Guide to **Planetary Nebulae**

Planetary nebulae have little to do with planets – they are small, typically rather spherical clouds of gas in deep space that can look a bit like planets when viewed through a small telescope. Often rather ghostly in appearance, they are formed when Sun-like stars reach a stage in their lifecycles where they shed their outer layers.

## Formation of a planetary nebula

As a star of similar mass to the Sun uses up the fuel it needs to create energy, it swells into a red giant. The red giant eventually ejects most of its material as a halo of gas – a planetary nebula. The rest of it collapses to form a hot dense star called a white dwarf.

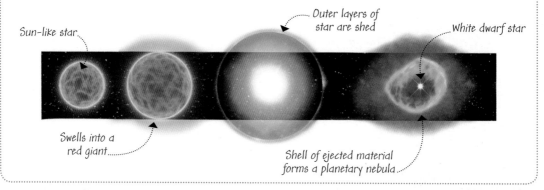

### Cat's Eye Nebula

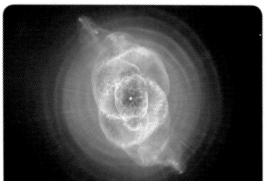

This nebula, in the constellation of Draco, is relatively bright and visible with a small telescope. Large instruments reveal its complex structure of arc- and bubble-like features and the central white dwarf star.

### Owl Nebula

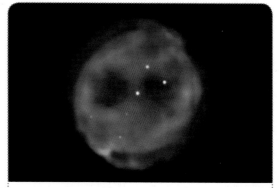

This nebula, also called M97, is a faint object in the constellation of Ursa Major, close to the star Merak (see p.166). A large telescope discerns the dark patches in the nebula that look like an owl's eyes.

# PLANETARY NEBULAE

**Kronberger 61 Nebula (in Cygnus)**

# KEY CONSTELLATIONS

## Observing **Vela**

The southern constellation of Vela sits partly in the Milky Way and contains several excellent objects for telescopic viewing. They include the Eight-Burst Nebula (a planetary nebula), a globular star cluster, and an open star cluster, as well as Regor, a bright binary star with additional fainter companion stars.

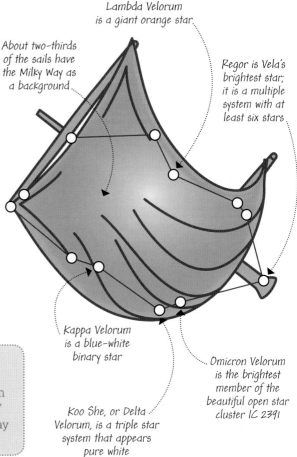

Lambda Velorum is a giant orange star

About two-thirds of the sails have the Milky Way as a background

Regor is Vela's brightest star; it is a multiple system with at least six stars

Kappa Velorum is a blue-white binary star

Koo She, or Delta Velorum, is a triple star system that appears pure white

Omicron Velorum is the brightest member of the beautiful open star cluster IC 2391

## **Vela, the Sails**

The constellation figure depicts sails that were originally part of the larger constellation of Argo Navis (see Carina, p.124). The objects in Vela recommended for telescopic viewing all lie close to the sails' outline. The Eight-Burst nebula appears as a pale oval disc with a tiny white dwarf star at its centre.

> **Remember** Two of Vela's stars – Kappa Velorum and Koo She (Delta Velorum) – are part of the False Cross asterism, which you may have explored in a "star hopping" exercise (see p.49). This is an excellent way to find Vela in the southern night sky.

## **Globular Star Cluster** NGC 3201

This star cluster is about 10 billion years old, lies 16,300 light years away, and contains hundreds of thousands of stars. Some can be seen individually through a telescope. Many are red giants, giving it a slightly orange hue.

# VELA

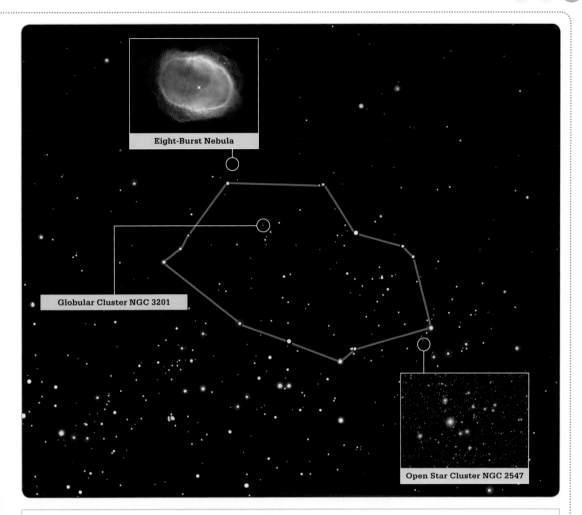

Eight-Burst Nebula

Globular Cluster NGC 3201

Open Star Cluster NGC 2547

## Southern hemisphere locators

From mid-latitudes (around 40°S) Vela is visible for part of the night in every month. View from January to March, when it is high in the south or nearly overhead at midnight, or from April to July in the south-western evening sky.

**1 October, 2am**
Looking southeast

**1 February, midnight**
Lookng high in south

**1 June, 10pm**
Looking southwest

# Safely observing **the Sun**

In addition to all the objects that can be observed at night, the centrepiece of our Solar System – the Sun – can be fascinating to study, but you must be stringent about safety when observing it. You should never look at the Sun with the naked eye, or through any instrument without using special filters.

## Projection techniques

The safest method to observe the Sun is to project its image onto a piece of white card. You can do this by directing the Sun's rays through a telescope eyepiece or one side of a pair of binoculars. The projection method is suitable for observing phenomena such as sunspots and solar eclipses; you do not look directly at the Sun so there is no chance of any light or other radiation impinging onto your eyes, which could cause severe damage, or even blindness. You may need to experiment a bit to get a clearly focused image on the card. Start by placing the card about 50cm (18in) from the telescope or binocular eyepiece. The further the card is from the eyepiece, the larger – but fainter – the image will be.

> **Remember** When projecting the Sun, be careful not to put your hand or anything flammable near the telescope or binocular eyepiece: the concentrated sunlight can cause a nasty burn or set something ablaze. Never leave a solar telescope – or a telescope with add-on filter – unattended, especially if children are present.

*Image of Sun focused on card*

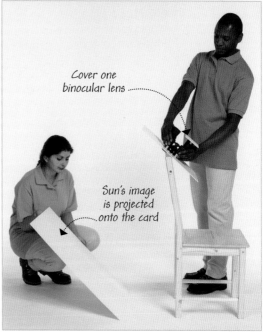

*Cover one binocular lens*

*Sun's image is projected onto the card*

## Telescope projection

Cap the main lens and finderscope. To aim at the Sun move the telescope around until you find the position where it casts the shortest shadow. Uncap the main lens and adjust the eyepiece to sharpen the image on the card.

## Binocular projection

Cover one binocular lens so that sunlight can only pass through the other side. Use the binocular lens to direct the Sun's image onto the card. Adjust the viewing card until the Sun's image is in focus.

# THE SUN

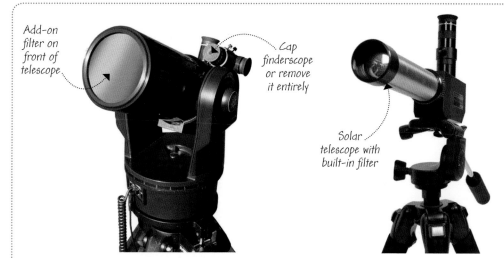

Add-on filter on front of telescope

Cap finderscope or remove it entirely

Solar telescope with built-in filter

## Solar filters and telescopes

An alternative to the projection method is to fit an ordinary telescope with a solar filter. Always adhere to the manufacturers' instructions, and only use add-on filters with the telescopes for which they are designed. Before use, hold the filter up and check that no pinpoints of light penetrate through scratches in its surface. If there are any flaws, reject it. Make sure the filter fits snugly over the telescope, and cannot be dislodged. Cap or remove the finderscope, then target the Sun using the "shortest shadow" method. Purpose-designed solar telescopes incorporating filters and other devices for viewing the Sun are also available.

## Solar phenomena to observe

There are a number of phenomena that can be observed on or around the Sun using projection methods or filters:
**Solar eclipses** – see pp.90–91.
**Sunspots** – these are darker, cooler regions on the Sun's surface, caused by localized disturbances in the Sun's magnetic field.
**Faculae** – brighter, more active regions on the Sun's surface.
**Solar prominences** – huge loops of material from the Sun extending into its atmosphere.
**Coronal mass ejections** – huge clumps of matter from the Sun that have broken off and are heading into space.
**Transit of Mercury** – when Mercury passes across the face of the Sun. The next transits occur on 9 May 2016 and 11 November 2019.

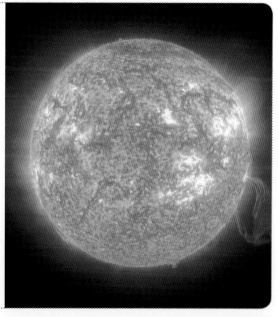

# A Quick Guide to **the Sun**

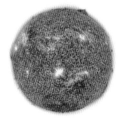

The Sun, our local star, is an incandescent ball of extremely hot plasma (ionized gas) with 750 times the mass of all the Solar System's planets combined. At its centre, vast amounts of energy are produced – this eventually escapes from the Sun's surface as heat, light, and other forms of radiation such as ultraviolet rays.

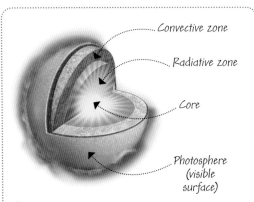

## Structure

The Sun has an extremely hot core where energy is produced by nuclear fusion. It is surrounded by a zone where energy is radiated outwards, then a convective zone where hot gas bubbles rise to the surface.

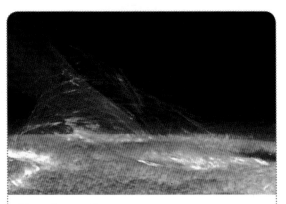

## Surface features

Various disturbances occur at the Sun's surface including violent explosions known as solar flares, jets of hot gas or huge loops of plasma shooting up or drifting into the Sun's atmosphere, and solar quakes.

## Sunspot cycle

Sunspots are dark, relatively cool regions that appear where hot gas cannot reach the Sun's surface. They emerge in bands either side of the Sun's equator. The number and pattern of spots varies over approximately 11-year cycles. In each cycle, the bands where spots appear gradually migrate towards the equator. At what are called solar maxima, when there are many spots near the equator, other types of surface disturbances occur. Some of these release vast amounts of energy and electrically charged matter into space.

**Year 1**

**Year 4**

**Year 7**

**Year 10**

**Year 12**

# THE SUN

## Sun profile

With a diameter of 1.4 million km (869,900 miles) and a mass 330,000 times that of Earth, the Sun is a stable, medium-sized yellow star.

**Distance from centre of Milky Way**
About 27,000 light years

**Orbits the galactic centre every**
225–250 million years

**Rotates once every**
25 days (equatorial region)
34 days (polar region)

**Greatest apparent magnitude**
-26.7 (see p.51)

**Nearest to Earth**
147,166,000km (91,445,000 miles)

**Greatest angular size in sky**
32.7 arcminutes (see p.12)

**Age**
4.6 billion years

**Number of planets** 8

**Size comparison**

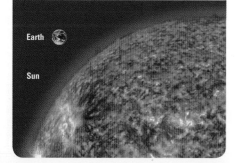

Earth
Sun

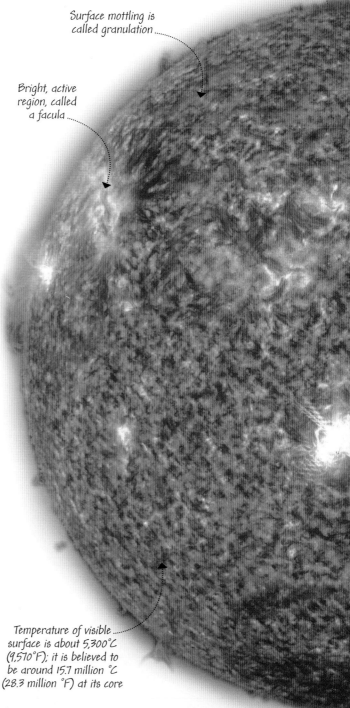

Surface mottling is called granulation

Bright, active region, called a facula

Temperature of visible surface is about 5,300°C (9,570°F); it is believed to be around 15.7 million °C (28.3 million °F) at its core

163

# KEY CONSTELLATIONS

## Observing **Virgo**

What the large zodiacal constellation of Virgo lacks in bright star clusters and nebulae, it makes up for in the number of the galaxies it contains – and some of them can be seen with a small telescope. Virgo straddles the celestial equator and can be seen equally well from both hemispheres.

**Remember** Virgo is most easily observed in the months of March, April, and May. It is difficult or impossible to view the constellation during September and October because it is too close to the Sun.

### Virgo, the Virgin

The constellation figure depicts a maiden goddess, identified by some as Demeter. In a limited area of the constellation – 50 million light years away but still worth targeting with a telescope – lies the Virgo cluster of galaxies. Highlights in other regions of this constellation include the Sombrero spiral galaxy (named for its resemblance to a Mexican hat) and the binary star Porrima.

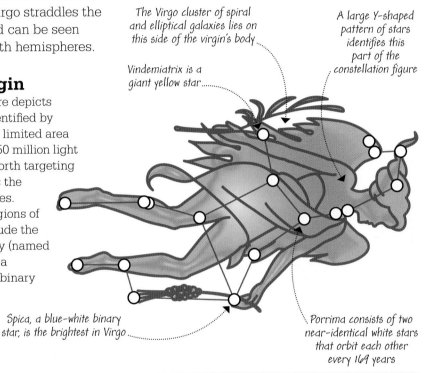

*The Virgo cluster of spiral and elliptical galaxies lies on this side of the virgin's body*

*Vindemiatrix is a giant yellow star*

*A large Y-shaped pattern of stars identifies this part of the constellation figure*

*Spica, a blue-white binary star, is the brightest in Virgo*

*Porrima consists of two near-identical white stars that orbit each other every 169 years*

## **Northern hemisphere** locators

From mid-latitudes (around 40°N) the constellation becomes visible from about early January in the southeastern sky – but only in the early morning. By April Virgo is prominent in the south at around midnight.

**1 January, 3am**
Looking southeast

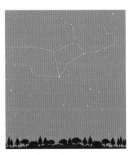

**1 April, midnight**
Looking south

**1 July, 9pm**
Looking southwest

# VIRGO

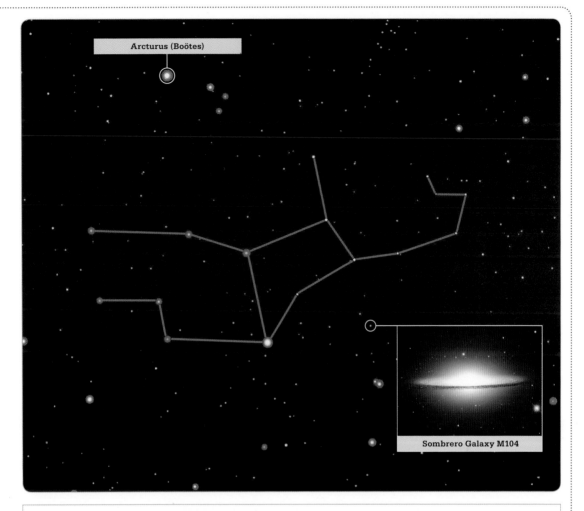

## Southern hemisphere locators

From mid-latitudes (around 40°S) Virgo is visible in the northeast early in the morning in January, or in the northern sky at around midnight in March and April. By July the constellation is in the northwest earlier in the evening.

**1 January, 3am**
Looking northeast

**1 April, midnight**
Looking north

**1 July, 9pm**
Looking northwest

## KEY CONSTELLATIONS

# Observing **Ursa Major**

A large northern constellation, Ursa Major is best known as the home of the Plough asterism, which provides a useful signpost to several stars and constellations (see pp.46–47). A lesser known fact is that within the constellation's borders lie some nearby galaxies that may be viewed through an amateur telescope.

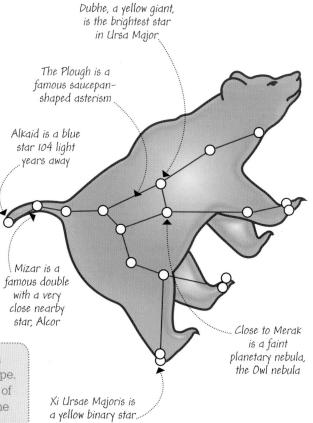

*Dubhe, a yellow giant, is the brightest star in Ursa Major.*

*The Plough is a famous saucepan-shaped asterism.*

*Alkaid is a blue star 104 light years away.*

*Mizar is a famous double with a very close nearby star, Alcor.*

*Xi Ursae Majoris is a yellow binary star.*

*Close to Merak is a faint planetary nebula, the Owl nebula.*

## Ursa Major, the Great Bear

The constellation figure is an adult bear striding across the northern sky. Try training your telescope on the Pinwheel Galaxy, located near the handle of the Plough, and Bode's Galaxy, nearer the north celestial pole. Also worth viewing are two stars very close to each other in the Plough's handle, Mizar, and Alcor.

> **Remember** Bode's Galaxy is one of Ursa Major's best objects to see with a telescope. This large spiral galaxy lies in the region of the bear's head. About one diameter of the Moon away from it is another galaxy that lies edge-on – known as the Cigar Galaxy.

## **Northern hemisphere** locators

For observers at mid-latitudes (around 40°N), Ursa Major is wholly visible at some point during the night all year round. Summer is a good season to view it, when it is prominent in the northwest in the late evening.

**1 December, 3am**
Looking northeast

**1 April, midnight**
Looking high in north

**1 August, 9pm**
Looking northwest

# URSA MAJOR

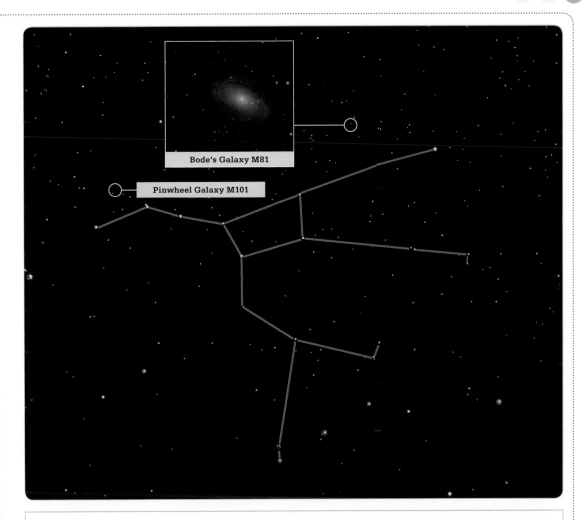

## Pinwheel Galaxy M101

Around 21 million light years away, the Pinwheel Galaxy (M101) forms the tip of a triangle with Mizar and Alkaid. It is face-on to Earth and just visible with a small telescope; a larger instrument will reveal the spiral arms.

KEY CONSTELLATIONS

# Observing **Hydra**

The large constellation of Hydra snakes more than a quarter of the way across the sky. Several different object types can be spied through a telescope in this constellation, including a planetary nebula, a spiral galaxy, and an open star cluster.

**Remember** One of the finest objects to view is star cluster M48, a loose collection of at least 80 stars occupying an area larger than that of the full Moon. Through a telescope these stars can be seen to be blue in colour, with some yellow ones mixed in.

## Hydra, the Water Snake

Hydra represents the monster that was killed by Hercules in the second of his labours. In the sky, it consists of a small group of stars representing the snake's head, and a dozen further stars that weave to the south of the constellations of Leo and Virgo. The main features of interest for telescopic viewing lie beneath the snake's head, tail, and belly.

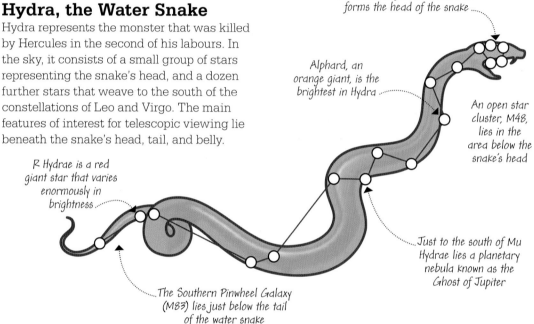

A distinctive five-star asterism forms the head of the snake

Alphard, an orange giant, is the brightest in Hydra

An open star cluster, M48, lies in the area below the snake's head

R Hydrae is a red giant star that varies enormously in brightness

Just to the south of Mu Hydrae lies a planetary nebula known as the Ghost of Jupiter

The Southern Pinwheel Galaxy (M83) lies just below the tail of the water snake

## **Northern hemisphere** locators

Hydra is so big that viewing the whole constellation at one time is difficult. For observers at mid-latitudes (around 40°N) however, it can be seen around midnight in early spring, looking from south to west.

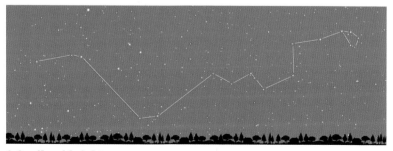

**1 April, midnight**

Looking south     Looking southwest     Looking west

# HYDRA

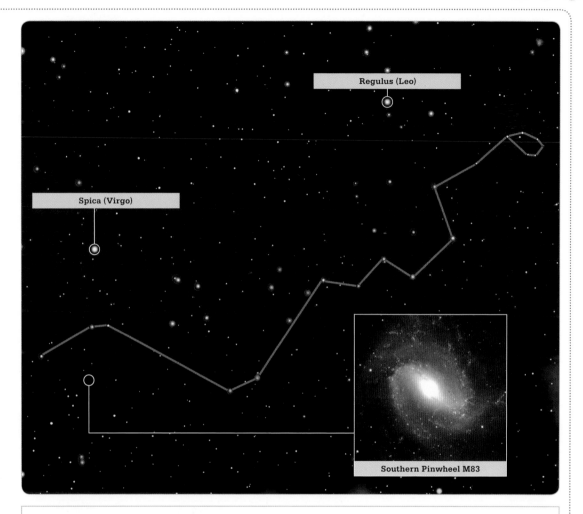

Regulus (Leo)

Spica (Virgo)

Southern Pinwheel M83

## Southern hemisphere locators

From mid-latitudes (around 40°S) it should be possible to view the whole constellation of Hydra at around 10pm in mid-February, looking in an area of the sky extending from the north to the southeastern horizon.

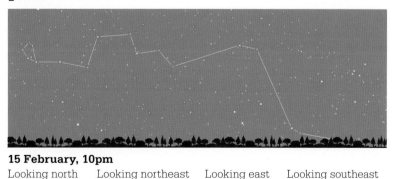

**15 February, 10pm**

Looking north    Looking northeast    Looking east    Looking southeast

OBSERVATION TECHNIQUES

# Observing a **Supernova Remnant**

Through your telescope you may be able to observe the remains of a gigantic explosion in space, first seen on Earth nearly 1,000 years ago. The Crab Nebula in Taurus is an example of a "supernova remnant".

## Locating the Crab Nebula

The best time to observe the Crab Nebula from either hemisphere is on evenings from December to March. First locate Taurus, just using your eyes (see pp.100–101 to remind yourself how to find it). Next, still with the naked eye, locate the two bright stars at the end of the bull's "horns". The one you are interested in is the star at the tip of the bull's right horn, called Zeta Tauri. Aim your telescope at this star. In the region around it, you should spot a small oval smudge in the sky, with a shape vaguely resembling a crab's shell: this is the Crab Nebula.

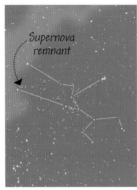

**Taurus from the northern hemisphere**

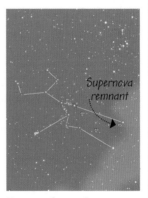

**Taurus from the southern hemisphere**

## Remnant in Vela

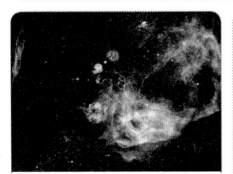

Other supernova remnants are dotted round the sky. This one, viewable in the southern sky only with a powerful telescope, is the remains of a star that exploded about 12,000 years ago.

## A cloud of debris

The Crab Nebula is the remains of a star that exploded some 7,500 years ago. As the star was about 6,500 light years away, the blast was not seen on Earth until 1054, when Chinese astronomers noted a new bright "star". The remnants are still expanding at the rate of about 1,500km per second, and the cloud is now about 11 light years across. At its centre is a strange object called a pulsar – an exceedingly dense star (a neutron star) that spins rapidly and emits powerful pulses of radiation, including X-rays and radio waves.

# SUPERNOVA REMNANT

**Crab Nebula M1 through the Hubble Telescope**

# A Quick Guide to **Supernovas**

A supernova is the exceedingly violent explosion of a high-mass star at the end of its life. It produces a vast outpouring of radiation, and a bright new light remains in the sky for weeks or months. Supernovas in our own galaxy are rare – the last conspicuous one was observed in 1604.

## How supernovas develop

When high-mass stars that are at least 8 times more massive than the Sun run out of fuel, they swell into supergiant stars. Eventually these explode as supernovas and – depending on their original mass – end up as either ultra-dense neutron stars or as black holes.

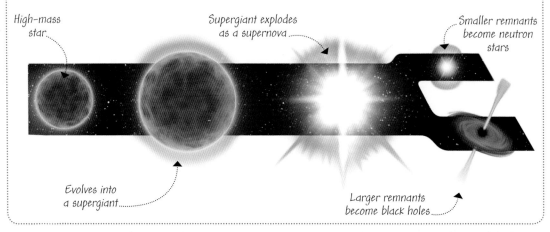

High-mass star

Supergiant explodes as a supernova

Smaller remnants become neutron stars

Evolves into a supergiant

Larger remnants become black holes

### Appearance of a supernova
The most recent supernova visible to the naked eye appeared in 1987 in a nebula outside our galaxy. The images above show how the nebula looked before (left) and after (right) the supernova came into view.

### Tycho's supernova remnant
The small red cloud at the top left of this image is the remnant of a supernova first seen on Earth in 1572. Danish astronomer Tycho Brahe noted that the supernova was so bright it was visible in broad daylight.

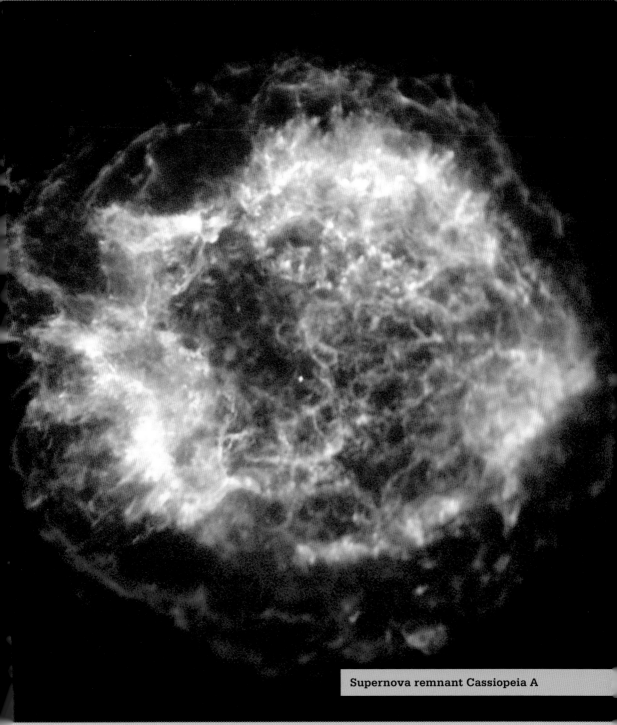

Supernova remnant Cassiopeia A

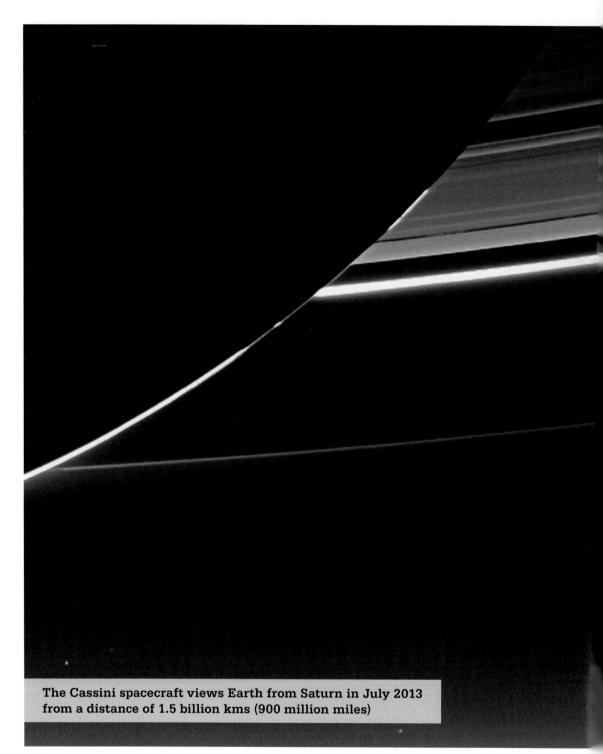

The Cassini spacecraft views Earth from Saturn in July 2013 from a distance of 1.5 billion kms (900 million miles)

SKY MAPS

# North Polar Sky

This region of the celestial sphere is centred on the North Celestial Pole, which is situated close to the prominent star Polaris. From Europe and North America much of this area of sky is always above the horizon. But from countries at mid-latitudes in the southern hemisphere – such as Australia – most of it is permanently hidden from view.

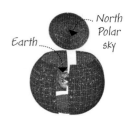

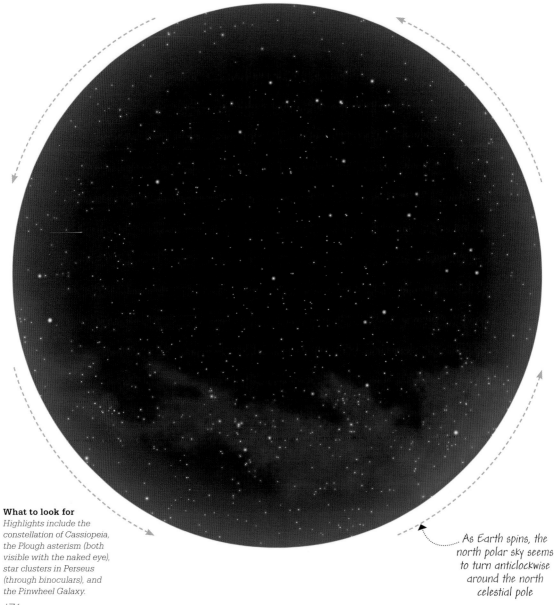

**What to look for**
*Highlights include the constellation of Cassiopeia, the Plough asterism (both visible with the naked eye), star clusters in Perseus (through binoculars), and the Pinwheel Galaxy.*

As Earth spins, the north polar sky seems to turn anticlockwise around the north celestial pole

# NORTH POLAR SKY

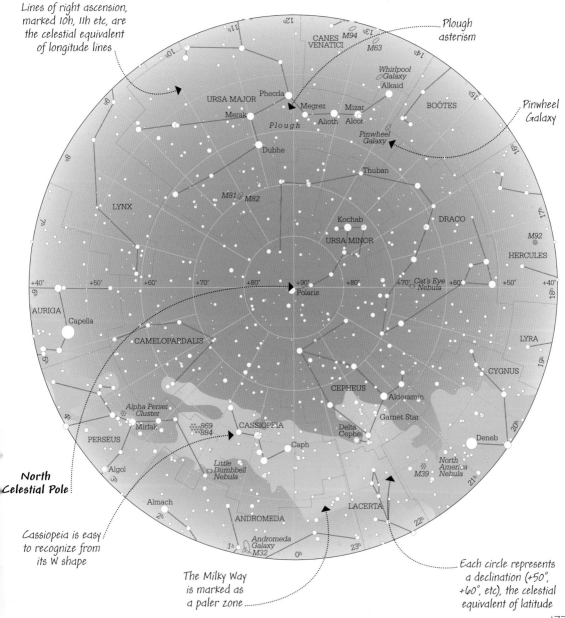

SKY MAPS

# South Polar Sky

This region of the celestial sphere is centred on the South Celestial Pole, which is situated in a fairly empty area of sky. From areas such as Australia, New Zealand, and southern Chile, much of this region of sky is always above the horizon. But from countries at mid-latitudes in the northern hemisphere, most of it is permanently hidden from view.

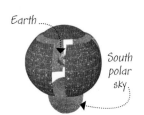

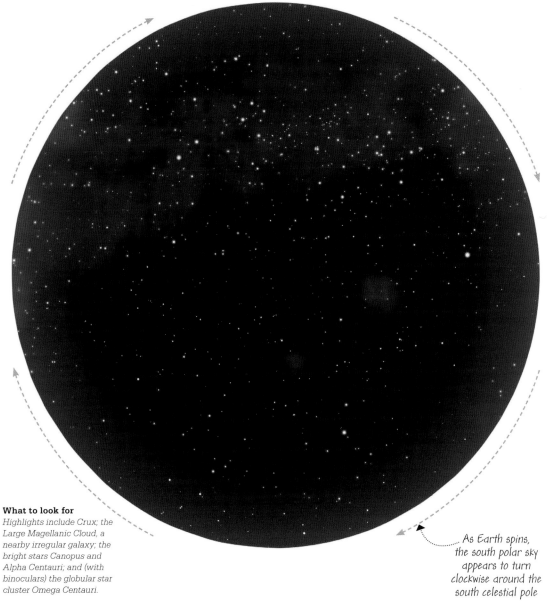

As Earth spins, the south polar sky appears to turn clockwise around the south celestial pole

**What to look for**
*Highlights include Crux; the Large Magellanic Cloud, a nearby irregular galaxy; the bright stars Canopus and Alpha Centauri; and (with binoculars) the globular star cluster Omega Centauri.*

# SOUTH POLAR SKY

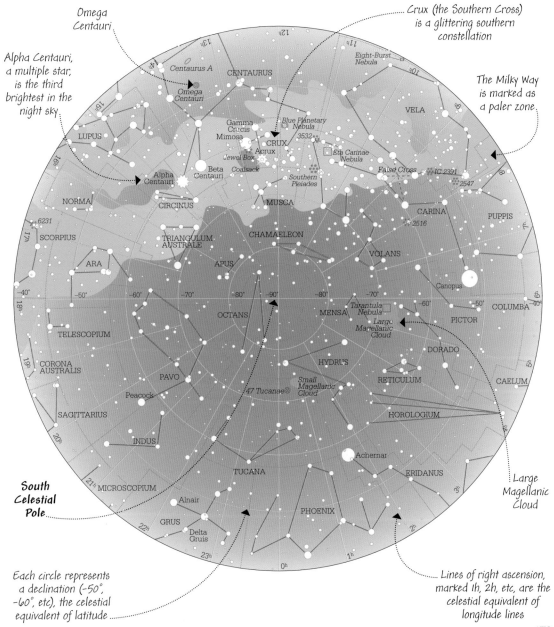

SKY MAPS

# Equatorial Sky 1

This part of the sky can be viewed during evenings from around September to November, looking to the south and overhead from the northern hemisphere or to the north (and overhead) from the southern hemisphere. Alternatively it can be viewed in the early hours of the morning, looking in the same directions, from around June to August.

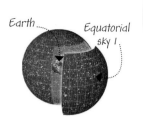

**What to look for**
*Highlights include the Square of Pegasus, as well as the Andromeda Galaxy and Helix Nebula, which are best viewed through binoculars or a telescope.*

# EQUATORIAL SKY 1

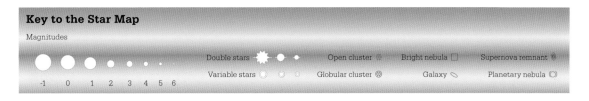
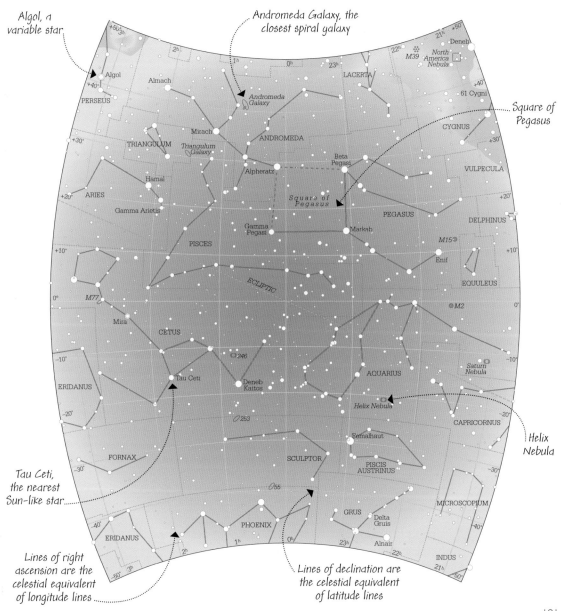

# Equatorial Sky 2

This sky region is best viewed on evenings from June to August, looking predominantly to the south from the northern hemisphere, or to the north from the southern hemisphere. It can also be viewed in the early hours of the morning from around March to May. The brightest part of the Milky Way dominates this part of the night sky.

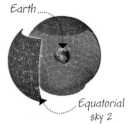

**What to look for**
*Highlights include Cygnus and Scorpius, and (with binoculars or a telescope) the Hercules Globular Cluster, and star clusters around Sagittarius.*

# EQUATORIAL SKY 2

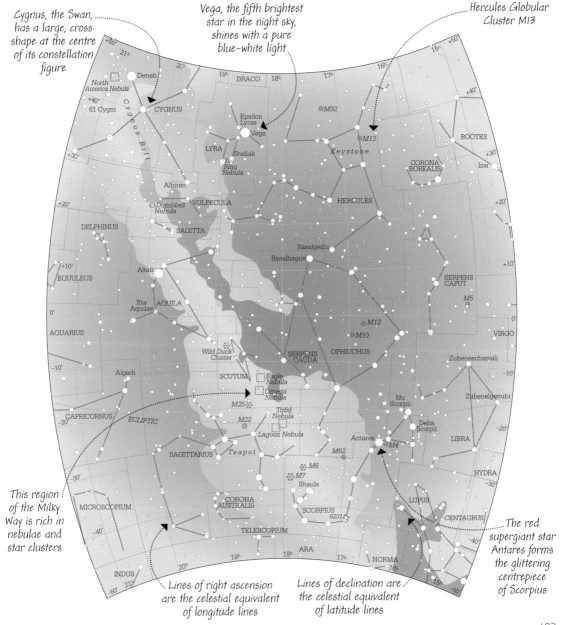

SKY MAPS

# Equatorial Sky 3

This area of sky can be seen during evenings from March to May, looking mainly to the south in the northern hemisphere, or to the north in the southern hemisphere. It can also be viewed in the hours after midnight from December to February. In this sky sector, there are only a few bright stars but many interesting deep-sky objects.

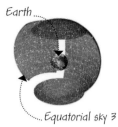

**What to look for**
*Highlights include Leo, stars Arcturus, Regulus, and Spica, and (with binoculars or a telescope) the Eight-Burst Nebula and the Southern Pinwheel Galaxy.*

# EQUATORIAL SKY 3

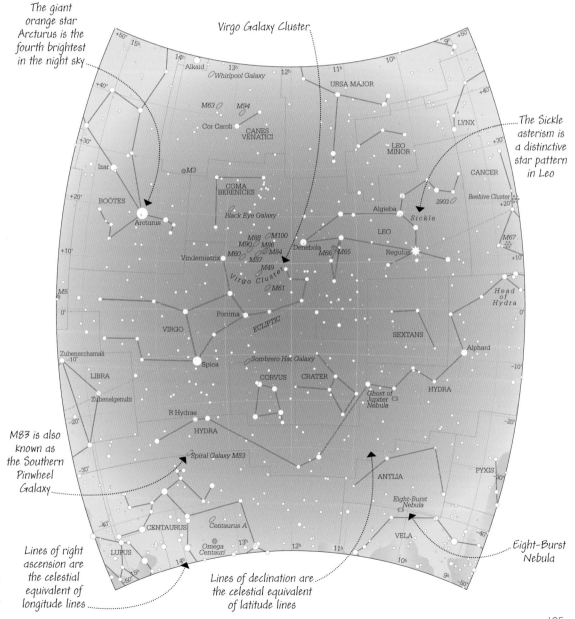

# Equatorial Sky 4

This part of the sky is best viewed during evenings from December to February, looking predominantly to the south in the northern hemisphere, or to the north in the southern hemisphere. It is also visible in the early mornings from September to November. Many brilliant stars and prominent constellations feature in this sector.

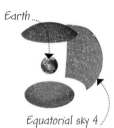

**What to look for**
*Highlights include Orion, Taurus, and Canis Major, and – through binoculars or a telescope – the Orion Nebula, and Pleiades, Hyades, and Beehive clusters.*

# EQUATORIAL SKY 4

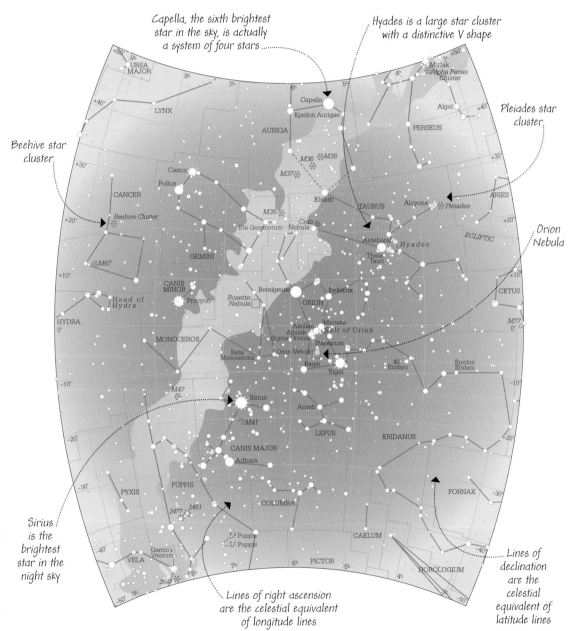

# Index

## A
Acrux 45, 49, 120
active galaxies 131
Aldebaran 61, 62, 63, 100
Algol 72, 181
Alkaid 166, 167
Alpha Centauri (Rigil
    Kentaurus) 45, 48, 49, 51,
    108, 121, 178, 179
Alpheratz 126, 128
Altair 37, 78, 79, 80, 81
Andromeda 126–127
Andromeda Galaxy (M31) 11,
    93, 126, 127, 128, 180, 181
angular diameter (size) 12
annular solar eclipses 90, 91
Antares 64, 65, 119, 183
aperture 94, 146
apparent magnitude 51
"apps" 23, 43, 58, 110, 148, 191
Aquarius 41, 67, 154–155
Aquila 37, 41, 78, 79
Arcturus 46, 165, 184
artificial satellites 8, 86–87
    Hubble Space Telescope 20,
        86, 87, 115, 171
    International Space Station
        16, 86, 87
asterisms 37, 49, 80, 168
    False Cross 44, 49, 158
    Keystone 80, 81, 107
    Orion's Belt 37, 60, 61, 62, 63
    Plough 37, 46–47, 132, 166,
        176, 177
    Sickle 37, 132, 133, 185
    Square of Pegasus 126, 127,
        128, 180, 181
    Teapot 37, 122, 123
asteroid belt 10, 14, 15
asteroids 10, 14, 15, 76
    impacts 112, 138
    see also meteors
astrology 67
astronomical aids 40, 58–59
atmospheres
    Earth 8, 12, 33, 150
    gas giants 14
    Jupiter 14, 70, 71
    Mars 112
    Mercury 12

Moon 98
Saturn 150
Titan 150
Uranus 15
Venus 11, 12, 84, 85
aurorae 8

## B
Beehive star cluster 186, 187
Bellatrix 62, 63
Beta Centauri (Hadar) 45, 48,
    49, 108
Beta Crucis 48, 49
Betelgeuse 60, 62, 63, 101
binary stars 52, 53, 64, 132,
    134, 164, 166
    eclipsing 53, 72
binoculars 16, 21, 23, 91, 93
    choosing and using 94–95
    projection through 160
black holes 21, 56, 78, 108, 131,
    172
Bode's Galaxy 166, 167
Boötes 46, 54, 165
Brahe, Tycho 172
brightness 36, 50, 51, 53
Bubble Nebula (NGC 7635) 117

## C
Callisto 152, 153
Caloris Basin 138
Canis Major 16–17, 62, 186, 187
Canis Minor 62
Canopus 51, 124, 125, 134, 135,
    178, 179
Capella 187
Carina 124–125, 135
Cassini spacecraft 174
Cassiopeia 54, 104–105, 176
Cassiopeia A supernova
    remnant 173
catadioptric telescopes 146
Cat's Eye Nebula 156
celestial equator 34
celestial poles 34, 35, 104, 134,
    147, 176, 177
    locating 42–45
celestial sphere 34–35, 38–45
Centaurus 48, 49, 108–109
Centaurus A galaxy
    (NGC 5128) 108, 109
Cigar Galaxy (M82) 21, 166

Coalsack Nebula 120, 121
coma 141, 142
comets 10, 18, 19, 76, 89,
    140–143
compass direction 58
compasses 23, 43, 45
conjunction 69, 82, 111, 136, 149
conjunctions, planetary 19
constellation figures 36–37
constellations 7, 36–37, 54
    locating 39, 58–59
    for meteor showers 74–75
    zodiac 66–69, 110–111, 148
Crab Nebula (M1) 7, 145, 170
craters 76, 112, 138, 139
    lunar 96, 97
Crux (Southern Cross) 25,
    44–45, 54, 120–121, 178
    starhopping from 48–49
Cygnus 41, 78–79, 157, 182, 183

## D
dark matter 21
dark nebulae 55, 116, 120, 121
date 40, 43, 45, 58
Delta Aquarii 74, 154
Delta Crucis 48, 49, 120
Delta Cygni (Rukh) 78, 80, 81
Deneb 78, 79, 80, 81
distances, light years 9
"Dog Star" see Sirius
Dorado 134–135
Draco 80
Dubhe 43, 46, 166
dwarf planets 14, 15

## E
Eagle Nebula (M16) 118, 119
Earth 9, 10, 11, 71, 152, 174
    atmosphere 8, 12, 33, 150
    distance from Sun 11, 163
    orbit 40–41, 67, 69, 76
    rotation 26, 27, 28, 38, 42, 44
    see also eclipses
eccentricity 89
eclipses 18, 25
    lunar 19, 32–33
    solar 19, 90–91, 160, 161
eclipsing binary stars 53, 72
ecliptic 66
Eight-Burst Nebula 158, 159,
    184, 185

emission nebulae 79, 116, 122,
    123
equator 35, 38, 42, 86
Equatorial sky 180–187
equipment 22–23, 58–59
    see also binoculars;
    telescopes
Eta Carinae Nebula
    (NGC 3372) 124, 125
Europa 152, 153
exoplanets 21
eyepieces 94, 95, 146, 160

## F G
False Cross 44, 49, 158
fields of view 147
filters 160, 161
finders 147
finderscopes 146, 147, 161
Gacrux 45, 120
galaxies 9, 16, 21, 36, 50, 64,
    88, 130–131
    Andromeda Galaxy (M31)
        11, 93, 126, 127, 128–129,
        180, 181
    Bode's Galaxy 166, 167
    Centaurus A Galaxy
        (NGC 5128) 108, 109
    Cigar Galaxy (M82) 21, 166
    M65 133
    M66 133
    NGC 4414 spiral galaxy 131
    Pinwheel Galaxy (M101)
        166, 167, 177
    Sombrero Galaxy (M104)
        164, 165
    Southern Pinwheel Galaxy
        (M83) 168, 169, 184, 185
    spiral 11, 57, 126, 127,
        128–129, 130, 131, 167, 168,
        169, 180, 181
    see also Milky Way
Galilean Moons 152–153
Ganymede 51, 152, 153
Garden Sprinkler Cluster
    (NGC 2516) 124, 125
gas giants 10, 14–15
    see also Jupiter; Neptune;
    Saturn; Uranus
Gemini 17, 41, 67
Geminids meteors 75, 77
Ghost of Jupiter 168

188

giant stars 145
  blue 50, 72, 100
  blue-white 104, 132
  orange 104, 168, 185
  red 50, 72, 100, 118, 126, 154, 156, 158, 168
  white 118
  yellow 78, 164
Globular Star Cluster (NGC 3201) 158, 159
globular star clusters 102, 106, 107, 118, 154, 155, 158, 159, 178, 182, 183
"Go-to" telescopes 147
gravity 88–89, 102, 131
  Jupiter 70
  Moon 98
"Great" comets 140, 141, 143
Great Nebula 124
Great Red Spot 14, 70, 152

## H

Hadar (Beta Centauri) 45, 48, 49, 108
Hale-Bopp, Comet 141
Halley's Comet 140
Heart Nebula (IC 1805) 105
Helix Nebula (NGC 7293) 154, 155, 180, 181
Hercules 80, 81, 106, 182, 183
Hercules Globular Cluster (M13) 106, 107, 182, 183
horizon 22, 41
Hubble Space Telescope (HST) 20, 86, 87, 115, 171
Hyades 16, 17, 100, 186, 187
Hydra 16, 168–169

## I J K

IC 1805 Heart Nebula 105
IC (Index Catalogue) numbers 95
impact craters 76, 96, 97, 112, 138, 139
International Space Station (ISS) 16, 86, 87
Io 152, 153
iridium flares 86
Jewel Box star cluster 120, 121
Jupiter 10, 14, 15, 16, 17, 25, 68–71
  apparent magnitude 51, 70

conjunctions 19, 69
Great Red Spot 14, 70, 152
moons 20, 51, 70, 71, 145, 146, 152–153
Keystone 80, 81, 107
Kronberger 61 Nebula 157
Kuiper Belt 10

## L

Lagoon Nebula (M8) 122, 123
Large Magellanic Cloud (LMC) 134, 135, 178, 179
latitude 38–39, 40, 41, 42, 43, 45, 58, 59
lenses 94, 146
Leo 37, 41, 46, 67, 132–133, 168, 169, 184, 185
Leo Minor 46, 47
Leonids meteor shower 132
light pollution 22, 54, 74, 128
light years 9
longitude 38
lunar cycle 28–31, 33, 74, 96
lunar diaries 26, 27
lunar eclipses 19, 32–33
Lynx 46
Lyra 78, 79, 107

## M

M1 Crab Nebula 7, 145, 170–171
M2 star cluster 154, 155
M8 Lagoon Nebula 122, 123
M10 star cluster 118
M12 star cluster 118
M13 Great Globular Cluster 106, 107, 182, 183
M16 Eagle Nebula 118, 119
M17 Omega Nebula 123
M31 Andromeda Galaxy 93, 126, 127, 128–129, 180, 181
M35 star cluster 103
M42 Orion Nebula 17, 52, 60, 93, 114–115, 186, 187
M45 Pleiades 16, 17, 63, 92–93, 100, 101, 186, 187
M48 open star cluster 168
M52 star cluster 105
M65 galaxy 133
M66 galaxy 133
M82 Cigar Galaxy 21, 166
M83 Southern Pinwheel Galaxy 168, 169, 184, 185

M97 Owl Nebula 156, 166
M101 Pinwheel Galaxy 166, 167, 177
M103 star cluster 104
M104 Sombrero Galaxy 164, 165
Maat Mons 84
McNaught, Comet 143
magnifying power 94, 146
magnitude 51
main asteroid belt 10, 14, 15
maps, sky 35, 54, 176–187
maria ("seas") 13, 93, 96, 97
Mars 10, 11, 13, 15, 19, 66, 67, 93, 110–113
Megrez 46, 47
Merak 43, 46, 47, 156, 166
Mercury 10, 11, 12, 19, 66, 93, 136–139, 161
Messier numbers 36, 95
meteor showers 18, 25, 74–75 76, 77, 100, 132, 154
meteorites 12, 76
meteors 8, 18, 74–75, 76, 141
Milky Way 9, 49, 50, 54–57, 64, 74–75, 118, 120, 122, 128, 134, 158
  nebulae 114, 116–117
  orbits in 88, 102, 163
  on sky maps 177, 179, 182, 183, 187
  star clusters 72, 102, 104, 108
  star-forming regions 114, 124, 125
  Sun's position in 11, 163
Mirach 126, 128
Mizar 166, 167
Moon 13, 15, 16, 24–25, 96–99, 128, 154, 168
  brightness 51, 68, 82, 99
  craters 93, 96, 97
  eclipses 19, 32–33
  maria 13, 93, 96, 97
  and meteor showers 74
  movements 26–27, 42, 44
  observing 20, 26–31, 93, 96
  orbit 27, 28, 88, 89, 98, 99
  phases 28–31, 33, 74, 96
  and solar eclipses 90–91
moons 20
  Jupiter 20, 51, 70, 71, 145, 146, 152–153

Neptune 20
Saturn 20, 150, 151
mounts for telescopes 147
multiple stars 53, 106, 126, 132
  see also Alpha Centauri; binary stars

## N

nebulae 18, 93, 102, 116–117, 122, 146, 154, 172, 183
  Cat's Eye Nebula 156
  Coalsack Nebula 120, 121
  Crab Nebula (M1) 7, 145, 170
  dark 55, 116, 120, 121
  Eagle Nebula (M16) 118, 119
  Eight-Burst Nebula 158, 159, 184, 185
  emission nebulae 79, 116, 122, 123
  Eta Carinae Nebula (NCG 3372) 124, 125
  Great Nebula 124
  Heart Nebula (IC 1805) 105
  Helix Nebula (NGC 7293) 154, 155, 180, 181
  Kronberger 61 Nebula 157
  Lagoon Nebula (M8) 122, 123
  NGC 1333 reflection nebula 73
  North America Nebula (NGC 7000) 78, 79
  Omega Nebula (M17) 123
  Orion Nebula (M42) 17, 52, 60, 93, 114–115, 186, 187
  Owl Nebula (M97) 156, 166
  planetary nebulae 145, 156–157, 158, 159, 168
  reflection nebulae 73, 116
  Saturn Nebula 154
  star-forming regions 114, 116, 124, 134, 144–145
  Tarantula Nebula (NGC 2070) 134, 135
Neptune 10, 11, 20, 66
neutron stars 172
NGC 752 Open Star Cluster 127
NGC 1333 reflection nebula 73
NGC 2070 Tarantula Nebula 134, 135
NGC 2158 star cluster 103
NGC 2516 Garden Sprinkler Cluster 124, 125

NGC 2547 Open Star Cluster 159
NGC 3201 Globular Cluster 158, 159
NGC 3372 Eta Carinae Nebula 124, 125
NGC 4414 spiral galaxy 131
NGC 5128 Centaurus A galaxy 108, 109
NGC 7000 North America Nebula 78, 79
NGC 7293 Helix Nebula 154, 155, 180, 181
NGC 7635 Bubble Nebula 117
NGC (New General Catalogue) numbers 36, 95
North America Nebula (NGC 7000) 78, 79
north celestial pole 34, 35, 42–43, 104, 176, 177
north polar sky 176–177
north polar star see Polaris
northern hemisphere 35, 41, 64, 108, 128, 176, 178
  equatorial sky 180–187
  Moon from 27
Nunki 80, 81

## O
Omega Centauri 108, 109, 178, 179
Omega Nebula (M17) 123
Oort Cloud 9, 10
Open Star Cluster M48 168
Open Star Cluster NCG 2547 159
Open Star Cluster NGC 752 127
open star clusters 62, 63, 102, 103, 127, 158, 159, 168
Ophiuchus 67, 107, 118–119
opposition 68, 69, 111, 149
optical doubles 52
orbits 89
  artificial satellites 86
  comets 89, 142
  Earth 40–41, 67, 69, 76
  Jupiter 68, 70, 71
  Mars 111, 112, 113
  Mercury 11, 136, 138, 139
  in Milky Way 88, 102, 163
  Moon 27, 28, 88, 89, 98, 99

Saturn 149, 150, 151
Solar System 10–11
Venus 82, 84, 85
Orion 16, 17, 25, 41, 60–61, 64, 186, 187
  Belt 37, 60, 61, 62, 63
  Nebula (M42) 17, 52, 60, 93, 114–115, 186, 187
  starhopping from 62–63
Orion Spur 56, 57
Owl Nebula (M97) 156, 166

## P
Pan-STARRS, Comet 142
partial eclipses
  lunar 32, 33
  solar 19, 90, 91
Pegasus 41
  Square of 126, 127, 128, 180
Perseids meteors 74–75, 76
Perseus 72–73, 176
phases of the Moon 28–31, 33, 74, 96
Phecda 46, 47
Pinwheel galaxy (M101) 166, 167, 177
planetary conjunctions 19
planetary nebulae 145, 156–157, 158, 159, 168
planets 9, 16, 88, 163
  gas giants 10, 14–15
  movements through zodiac 66–69, 110–111, 148–149
  rocky planets 10, 12–13
  see also Earth; Jupiter; Mars; Mercury; Neptune; Saturn; Uranus; Venus
planispheres 23, 40, 43, 58, 59, 74, 110
Pleiades (M45, Seven Sisters) 16, 17, 63, 92–93, 100, 101, 186, 187
Plough 37, 42–43, 166, 176, 177
  starhopping from 46–47, 132
Pluto 14
Polaris (north polar star) 42, 43, 44, 46, 51, 176
Porrima 164
Procyon 16, 62
projection techniques 160
Proxima Centauri 108
pulsars 170

## R
radiant 74, 75, 76
radiation 21, 78, 131, 142, 160, 170, 172
red dwarves 50
red giants 50, 72, 100, 118, 126, 154, 156, 158, 168
reflection nebulae 73, 116
reflector telescopes 146
refractor telescopes 146
Regulus 46, 132, 169, 184, 185
retrograde motion 67
Rigel 51, 60, 63
Rigil Kentaurus see Alpha Centauri
rocky planets 10, 12–13
  see also Earth; Mars; Mercury; Venus
Rukh (Delta Cygni) 78, 80, 81

## S
safety and the Sun 91, 160–161
Sagittarius 37, 67, 80, 81, 122
Sagittarius A 56
satellites see artificial satellites; moons
Saturn 10, 11, 14, 15, 19, 66, 148–151, 174
  moons 20, 150, 151
  ring system 11, 15, 145, 146, 148, 150, 151
Saturn Nebula 154
Scorpius 64–65, 182, 183
"seas" (maria) 13, 93, 96, 97
seasons 40–41, 43, 45
Serpens Caput 118, 119
Serpens Cauda 118, 119
shooting stars see meteors
Sickle 37, 132, 133, 185
Sirius ("Dog Star") 16, 17, 51, 61, 62, 187
size (angular diameter) 12
sky maps 176–187
solar eclipses 19, 90–91, 160
solar filters 160, 161
solar flares 162
solar mass 50
solar maxima 162
solar prominences 161
solar quakes 162
Solar System 9, 10–15, 56, 57, 76
  see also comets; planets; Sun

solar telescopes 160, 161
Sombrero Galaxy (M104) 164, 165
south celestial pole 34, 35, 44–45, 134, 178, 179
south polar sky 178–179
Southern Cross see Crux
southern hemisphere 35, 128, 176, 178
  equatorial sky 180–187
  Jupiter from 68
  Moon from 27, 31, 96
  Orion from 61, 114
  Scorpius from 64, 65
Southern Pinwheel Galaxy (M83) 168, 169, 184, 185
Southern Pleiades 124
Spica 19, 164, 184, 185
spiral galaxies 57, 130, 131, 167, 168, 169
  see also Andromeda Galaxy; Milky Way
Spitzer Telescope 73
Square of Pegasus 126, 127, 128, 180, 181
star clusters 50, 64, 72, 93, 102–103, 104, 182, 183
  Beehive 186, 187
  in Carina 124, 125
  Garden Sprinkler Cluster (NGC 2516) 124, 125
  globular 102, 106, 107, 118, 154, 155, 158, 159, 178, 182, 183
  Globular Cluster (NGC 3201) 158, 159
  Hercules Globular Cluster (M13) 106, 107, 182, 183
  Hyades 16, 17, 100, 186, 187
  Jewel Box 120, 121
  M2 154, 155
  M10 118
  M12 118
  M16 118, 119
  M35 103
  M48 Open Cluster 168
  M52 105
  M103 104
  NCG 752 Open Cluster 127
  NGC 2158 103
  NGC 2547 159
  Omega Centauri 108, 109, 178, 179

open 62, 63, 102, 103, 127, 158, 159, 168
in Perseus 72, 176
Pleiades (M45) 16, 17, 63, 92–93, 100, 101, 186, 187
in Sagittarius 122, 182
star systems 52–53, 78, 100, 106, 126, 132, 187
*see also* Alpha Centauri; binary stars
star-forming regions 104, 114, 116, 124, 134, 144–145
starhopping 25
from Crux 48–49
from Orion 62–63
from Plough 46–47, 132
from Summer Triangle 80–81
stars 36, 50–51, 88
apparent movement 34–35
giants 50, 72, 78, 100, 104, 118, 132, 145, 164, 185
lifecycles 50, 116, 134, 154, 156, 172–173
locating 58–59
multiple 53, 106, 126, 132, 178, 179
supergiants 72, 78, 120, 124, 172

variable 53, 72, 181
*see also* binary stars; Sun
Summer Triangle 78, 79
starhopping from 80–81
Sun 9, 10, 11, 40, 50, 52, 88, 162–163
brightness 13, 51, 60, 82, 163
and comets 142
distances of planets from 11, 70, 71, 85, 113, 139, 151, 163
eclipses 19, 25, 90–91, 160
ecliptic 66
and lunar eclipses 32
mass 50, 162, 172
observing safely 91, 160–161
and phases of the Moon 28, 29, 30, 31
transit of Mercury 161
transit of Venus 18
and zodiac 66
*see also* Solar System
sunspots 158, 160, 161
supergiants 72, 120, 172
*see also* Antares; Betelgeuse; Canopus
supernovas 145, 170–173
Swift-Tuttle, Comet 76

## T
Tarantula Nebula (NGC 2070) 134, 135
Tau Ceti 181
Taurids meteor shower 100
Taurus 61, 62, 63, 67, 100–101, 186, 187
Crab Nebula 7, 145, 170–171
*see also* Pleiades
Teapot 37, 122, 123
telescopes 20–21, 23, 91, 145
choosing and using 146–147
Hubble Space Telescope 20, 86, 87, 115, 171
projection with 160
solar filters 160, 161
Spitzer Telescope 73
time 40, 43, 45, 58
Titan 150
total eclipses
lunar 32, 33
solar 19, 90, 91
transits 18, 161
Trapezium Cluster 52
Triangulum Australe 28, 49
Tycho crater 96
Tycho's supernova remnant 145, 172

## U V
Universe 8, 9, 21, 88, 102, 130
Uranus 10, 15, 20, 66, 84
Ursa Major 156, 166–167
*see also* Plough
Ursa Minor 43, 54
Polaris 42, 43, 44, 46, 47
variable stars 53, 64
Algol 72, 181
*see also* binary stars
Vega 51, 78, 79, 80, 81, 107, 183
Vela 124, 158–159, 170
Venus 10, 11, 12, 25, 66, 82–85, 136, 137
brightness 51, 85
conjunctions 19
transit of 18
Virgo 41, 46, 67, 164, 168, 169

## W–Z
water 12, 13, 84, 112, 113
white dwarves 50, 156
zodiac 40, 41, 66–69, 110–111, 148–149
zodiac signs 67

## Useful resources

*Planispheres*
**DK Planisphere and Starfinder**, Carole Stott and Giles Sparrow. Latitude options: 55°N (UK); 35°N (US); 35°S (Australia)
**Philip's Planisphere,** Latitude options: 23.5°N, 32°N, 42°N, 51.5°N, 35°S
**The Night Sky Planisphere** David Chandler. Latitude options: 20–30°N, 30–40°N, 40–50°N, 50–60°N, 20–50°S
**Rob Walrecht Planisphere** Latitude options: from 40°S to 66.5°N and several languages
**Equatorial Guide to the Stars** (Ken Press/Ken Graun) Latitude options: 30°S-30°N

*Desktop/laptop Software*
**Starry Night Complete Space and Astronomy Package** (Simulation Curriculum Corp) Beginner level for PCs and AppleMacs
**Redshift 7 premium** (United Soft Media) Advanced level, for PCs only

*iPhone and iPad "apps"*
**Stellarium Mobile** (Noctua Software)
**Pocket Universe** (John Kennedy)
*Android "apps"*
**Mobile Observatory** (KreApp Development Software)
**Sky Map** (Google)

*Books*
**The Practical Astronomer**, Will Gater and Anton Vamplew (DK)
**Stars and Planets Nature Guide**, Robert Dinwiddie, Will Gater, Giles Sparrow, and Carole Stott (DK)
**Eyewitness Companions: Astronomy**, Ian Ridpath (DK)
**Space**, Carole Stott, Robert Dinwiddie, David Hughes, and Giles Sparrow (DK)
**Universe**, General Editor: Martin Rees (DK)
**Unfolding our Universe** Iain Nicolson (Cambridge University Press)

*Online tools and calendars*
**NASA Skywatch (for ISS, HST, and other satellites)** http://spaceflight.nasa.gov/realdata/sightings/SSapplications/Post/JavaSSOP/JavaSSOP.html
**Astronomical Calendar night sky for UK and northern Europe** http://astronomical-calendar.org.uk/
**Sea and sky** http://www.seasky.org/astronomy/astronomy-calendar-current.html
**Astronomy magazine** http://www.astronomy.com

# About the Author

**Robert Dinwiddie** is a science writer, editor, and educator who specialises in astronomy, cosmology, Earth science, and the history of science. A keen traveller and amateur astronomer, he has worked in a variety of media, including books, magazines, and electronic publishing. This has included managing the development of the DK CD-ROM, *Eyewitness Encyclopedia of Space and the Universe*, co-authoring DK astronomy encyclopedias *Universe* and *Space*, the *Stars and Planets* nature guide, further DK science-based encyclopedias including *Earth, Science, Ocean,* and *Violent Earth*, and numerous other books.

# Acknowledgements

Many people helped in the making of this book. DK would particularly like to thank:
**Additional writing** Ben Morgan; **Photography** Peter Anderson
**Editorial assistance and proofreading** May Corfield;
**Design assistance** Prashant Kumar, Clare Marshall;
**Illustrations** Vanessa Hamilton; **Index** Jane Coulter

## Picture credits

DK would like to recognise the efforts of all the enthusiastic astronomers and astro-photographers who share their images online. The publishers would like to thank the following for their kind permission to reproduce their photographs in this book:
(Key: a-above; b-below/bottom; c-centre; f-far; l-left; r-right; t-top)

**02 Courtesy Jim Misti. 4 NASA:** Rocky Raybell (b). **6 ESO:** B.Tafreshi (twanight.org). **7 NASA:** JPL-Caltech/R. Gehrz (University of Minnesota) (cr). **9 NASA:** JPL-Caltech, SDSS (tr). **10 NASA:** Solar & Heliospheric Observatory (l). **13 Dorling Kindersley:** NASA/digitaleye © Jamie Marshall (tr). **14 NASA:** ESA, Buie/Southwest Research Institute (t). **18 NASA:** Rocky Raybell. **19 ESO:** Y. Beletsky (b). **NASA and The Hubble Heritage Team (AURA/STScl):** D. Jewitt (UCLA) (tr). **20 Fraser Gunn:** (tr). **NASA and The Hubble Heritage Team (AURA/STScl):** John Spencer (Lowell Observatory) (tl); L. Sromovsky and P. Fry (University of Wisconsin), H. Hammel (Space Science Institute),K. Rages (SETI Institute) (bl). **NASA:** E. Karkoschka (University of Arizona) and H.B. Hammel (Space Science Institute, Boulder, Colorado) (br). **21 NASA:** ESA, and The Hubble Heritage Team (STScI). **33 David Cortner. 42 © Australian Astronomical Observatory/David Malin. 44 Dan Heller Photography. Chris Picking:** (tl, tr). **50 NASA and The Hubble Heritage Team (AURA/STScl):** Hubble SM4 ERO Team. **51 NASA:** STScI Digitized Sky Survey/Noel Carboni (r). **52 NASA and The Hubble Heritage Team (AURA/STScl):** C.R. O'Dell and S.K. Wong (Rice University) (b). **54 ESO:** J.Perez (tl). **55 ESO**. **56 NASA:** ESA, SSC, CXC, and STScI (l). **70 NASA:** JPL (br). **73 NASA:** JPL-Caltech/Harvard-Smithsonian CfA (b). **74 Amir Hossein Abolfath:** (l). **75 Jens Hackmann. 77 Arman Golestaneh. 79 NASA:** JPL-Caltech (b). **82–83 Peter Lawrence**. **86 ESO:** F.Char (br). **87 Mark Humpage. 90–91 Ben Cooper/LaunchPhotography.com. 91 Rainbow Symphony Inc:** (tr). **92–93 NASA and The Hubble Heritage Team (AURA/STScl). 96 ESA/Hubble:** (br). **NASA:** (bl). **97 NASA:** JPL/USGS. **101 NASA and The Hubble Heritage Team (AURA/STScl):** (inset). **102 Jean-Charles Cuillandre/Canada-France-Hawaii Telescope Corporation:** (br). **NASA:** F. R. Ferraro (ESO/Bologna Obs.), M. Shara (STSci/AMNH), Hubble Heritage Team (AURA/STScI/NASA) (bl). **103 Jean-Charles Cuillandre/Canada-France-Hawaii Telescope Corporation. 105 Courtesy Jim Misti:** (Inset). **NASA:** JPL-Caltech/UCLA (br). **107 Danny Lee Russell:** (t). **108 ESO:** Gendler, J.E. Ovaldsen & S. Guisard (b). **109 ESO:** (Inset). **112 NASA:** JPL/MSSS (br). **114 Stephen W. Ramsden/www.solarastronomy.org:** (bl). **115 NASA and The Hubble Heritage Team (AURA/STScl):** ESA, M. Robberto (STScIESA). **116 ESO:** (t, br). **NASA:** The Hubble Heritage Team (STScI) (bl). **117 Larry van Vleet. 119 ESO:** (Inset). **120 ESO:** S.Brunier (b). **121 ESO:** (t/Inset). **123 ESO:** INAF-VST/OmegaCAM. (t/Inset). **124 NASA. 127 David Davison:** (br). **128 Matt Kingsnorth:** (bl). **129 R Gendler:** Robert Gendler. **130 ESO:** (bl). **NASA:** Laurent Drissen, Jean-Rene Roy and Carmelle Robert (Department de Physique and Observatoire du mont Megantic, Universite Laval) (br); ESA and The Hubble Heritage Team (STScI/AURA) (tl); ESA, and The Hubble Heritage Team (STScIAURA) (tr). **131 NASA:** ESA/John Bahcall (Institute for Advanced Study, Princeton) Mike Disney (University of Wales) (b); The Hubble Heritage Team (STScI) (t). **133 David Ratledge:** (t/Inset). **134 Justin Whitten:** (b). **135 NASA:** JPL-Caltech (t/Inset). **137 ESO:** B.Tafreshi (twanight.org). **138 NASA:** Courtesy of Science/AAAS. **140 Fred Espenak. 141 Philipp Salzgeber:** (b). Doug Zubenel: (t). **142 NASA:** Astronomy Education Services/Gingin Observatory (bl). **143 ESO:** S. Deiries. **144-145 NASA:** JPL-Caltech/UCLA. **150 NASA:** JPL-Caltech/Space Science Institute (br). **153 Fraser Gunn:** (tr). **NASA:** NASA Planetary Photojournal (br). **155 Photo courtesy of Bryan Bradley, Babylon, NY, USA:** (cr/Inset). **NASA:** D. Williams, N. A. Sharp, AURA, NOAO, NSF (tl/Inset). **156 NASA and The Hubble Heritage Team (AURA/STScl):** (bl). Keith Quattrochi: (br). **157 Gemini Observatory:** AURA. **158 NASA:** The Hubble Heritage Team (STScI) (b). **159 ESO:** (cr/Inset). NASA: The Hubble Heritage Team (STScI)/AURA/ESA) (t/Inset). **161 NASA:** Solar & Heliospheric Observatory (br). **162 NASA:** (tl, cr). **163 NASA:** (bl, r). **165 NASA and The Hubble Heritage Team (AURA/STScl):** (cr/Inset). **167 NASA:** JPL-Caltech (br). Robert Vanderbei (t). **169 ESO:** (cr/Inset). **170 Chandra X-Ray Observatory:** NOAO/CTIO (cl). **171 NASA:** JPL-Caltech/R. Gehrz (University of Minnesota). **172 © Australian Astronomical Observatory/David Malin:** (bl). **NASA:** JPL-Caltech/UCLA (br), 173 Chandra X-Ray Observatory: NASA/CXC/MIT/UMass Amherst/M.D.Stage et al. **174-175 NASA:** JPL-Caltech/Space Science Institute.

Every effort has been made to trace the copyright holders. The publishers apologise for any unintentional omissions and would be pleased, in such cases, to place an acknowledgement in future editions of this book.

All other images © Dorling Kindersley
For further information see: **www.dkimages.com**